TELECOMMUNICATIONS IN CRISIS

D0929928

TELECOMMUNICATIONS IN CRISIS:
THE FIRST AMENDMENT, TECHNOLOGY, AND DEREGULATION

EDWIN DIAMOND AND
NORMAN SANDLER,
AND MILTON MUELLER

Library of Congress Cataloging in Publication Data

Diamond, Edwin.
    Telecommunications in crisis.

    Contents: The FCC and the deregulation of telecommunications
technology/Edwin Diamond and Norman Sandler—Reforming
telecommunications regulation/Milton Mueller.
    1. Telecommunication—Law and legislation—United States.
2. United States. Federal Communications Commission.
3. Broadcasting—Law and legislation—United States.   4. Freedom of
the press—United States.   5. Telecommunication policy—United
States.   I. Sandler, Norman.   II. Mueller, Milton.   III. Title.
KF2765.D5   1983   343.73'0994   83-15038
ISBN 0-932790-39-9   347.303994

Printed in the United States of America.

CATO INSTITUTE
224 Second Street SE
Washington, D.C. 20003

CONTENTS

FOREWORD

Telecommunications is one of today's most fascinating subjects because this field promises so much for our future. Communications technologies provide an impressive example of our nation's scientific achievements; new and more amazing devices seem to be invented on a daily basis. The businesses that spring from these technologies form a healthy and increasingly vital segment of our economy. And the public benefits through an expanding array of information and entertainment choices.

At the same time, the issues surrounding telecommunications are some of the most complicated and confusing ever faced by legislators, consumers, and entrepreneurs. The government struggles to deal with ever-evolving technologies. Businesses battle government restraints and competition from other businesses. And consumers are often overwhelmed by a flood of unfamiliar products, services, and companies.

The Cato Institute has put together a two-faceted examination of this subject to bring order where chaos seems to exist. The first half of the book consists of two journalists' look at the telecommunications world as it is. The second half is a policy analyst's look at that world as it should be.

Edwin Diamond and Norman Sandler, the journalists, outline the current telecommunications regulatory scheme and review the many problems that scheme has created. Their observations on the past and present struggles over radio, low-power television, cable, satellites, and cellular radio illustrate how the telecommunications universe is allowing new entrants and increasing competition.

Diamond and Sandler take special care to highlight the First Amendment ramifications of those new technologies. Few telecommunications issues are more important. Our democratic society requires that these modern technologies be incorporated into our fundamental tradition of freedom of expression. How that will be accomplished presents a challenge for our policymakers.

In a far less complex time in our history, our founders considered freedom of expression to be so vital that they prohibited the government from intruding into any of the then-available forms of communications—speech and press. Since the First Amendment was drafted, however, science has created methods of communication that were not foreseen by the founders, and our policies have not kept pace with these changes. The government *does* intrude deeply into radio, television, and cable, and our courts have upheld these intrusions. Government regulation of the content of these media is premised on the asserted "scarcity" of the electromagnetic spectrum.

Diamond and Sandler correctly point out that modern technical advancements have seriously eroded this scarcity rationale. There are now over 9,000 radio stations and 1,000 TV stations. The nation is rapidly being wired for cable systems capable of providing dozens of channels. Satellites will soon be beaming even more programming directly into our homes. These technologies and others, if we allow them to, can provide the diversity that never can be achieved by government regulation.

Diamond and Sandler suggest that this technological explosion will lead the Congress and the courts to loosen their grip on the electronic media and eliminate content regulations such as Equal Time and the Fairness Doctrine. While I hope the authors are correct, I am not so optimistic. It may be difficult to persuade the Congress to remove these regulations, many of which were written to benefit incumbent politicians.

Nonetheless, the fight is one that must be joined. Our First Amendment traditions are infringed by these regulations: They hamstring broadcasters, and they deprive the public of better and more diverse service.

Worst of all, these regulations have the potential to ensnare the printed press. New telecommunications technologies are merging the print and electronic media. Newspapers are increasingly dependent upon electronic methods of gathering and disseminating news. If we do not eliminate the laws that permit the government to oversee the conduct of the electronic media, these laws could be used to attack the print media's First Amendment freedoms.

I support an affirmative legislative effort to remove these regulations once and for all. We cannot rely only upon technological forces and the good will of Congress and the courts. Neither can we leave this necessary task to the Federal Communications Commission. It is bound by law to carry out statutory obligations, and while the current FCC is attempting to deregulate communications, a future Commission could lose interest in this important work or it could easily reimpose controls. The Congress itself must be motivated to act by an aware and concerned public.

Ultimately, I believe that we will want to write into our Constitution new protections for electronic expression. No other method of protecting our liberties is as secure. Any lesser protection for our fundamental rights disserves us all.

Part two of the book contains Milton Mueller's proposal for a free market in telecommunications. This market would be created by defining freely transferable property rights in the electromagnetic spectrum. Mueller asserts that the introduction of price mechanisms into the frequency allocation process would be a welcome change from the present inefficient and wasteful centralized method. Furthermore, he says, this new market would permit open entry into radio spectrum communications, thereby increasing competition among users.

Mueller is not the first to suggest that the radio spectrum could and should be governed by free market mechanisms. These proposals have been around since the 1950s. Mueller's provocative study builds upon these earlier ideas by refining the all-important definition of property rights in the spectrum.

Free market proposals, however, have not been implemented, nor have many policymakers seriously considered them.

The reason for that is apparent. Milton Mueller's proposal, like his predecessors', will be viewed as too far-reaching and impractical. A free market approach would radically revise our nation's telecommunications framework. Any attempt to establish such a system would almost certainly lead to political and economic battles that would make those outlined in part one of this book look mild by comparison.

Nonetheless, free market proposals should not be dismissed out of hand. The question is whether a free market approach would improve the system, which is fraught with the failures noted by Diamond and Sandler.

In conclusion, I hope that this book will stir further discussion about how we can make our telecommunications structure more open, more competitive, and more responsive to our nation's needs. Jefferson's dream of a free marketplace of ideas can become a reality only in a nation where freedom of expression is guaranteed.

SENATOR BOB PACKWOOD
Chairman
Senate Committee on Commerce,
Science and Transportation

THE FCC AND THE DEREGULATION OF TELECOMMUNICATIONS TECHNOLOGY

Edwin Diamond and
Norman Sandler

I. Deregulation and Unregulation

It is an axiom in Washington that American business is against government regulation—except those regulations that gore somebody else's ox. "Hands off," business shouts publicly to the regulators, and then in an off-stage whisper adds, "but keep your hands *on* my competitors and suppliers. . . ."

This doublespeak is particularly characteristic of the communications business, or as it has more grandly been christened, the telecommunications industry. Over the years, the established network broadcasters have fought for their own freedom from a variety of Federal Communications Commission regulations, such as FCC limits on station ownership, while lobbying the FCC to keep the rival, and growing, cable television competition tied up in knotty rules. The cable TV industry, in turn, has demanded its own freedom from government control while clinging to the protective rules that helped feather its nest, such as the regulations that exempt the "poor little" cable system operators (Time Inc., and Warner Communications, among others) from having to pay copyright royalties for many of the television programs they cablecast.

So it should come as no suprise to anyone that despite all the talk about the whirlwinds of change and deregulation sweeping through Ronald Reagan's Washington—forces actually set in motion years before, first in the Nixon-Ford era and later by the Carter administration—we are in fact getting not so much deregulation as something that might be called unregulation. This hybrid exists somewhere in the gray area between full control and no control. The new style is much like some "company" police force in, say, an energy boom town in Mexico or Montana, enforcing the law selectively or only when provoked. When applied to telecommunications and the emerging new media technologies, the new style means a world of mixed rule-making and marketplace freedoms, of continued "public interest" regulations and private enterprise initiatives.

This is a world that is not ideologically pure, either for 1960s Great Society Democrats—who believe, following a tradition dating back to Plato, that man must be made to be good—or for 1980 Reagan Repub-

3

licans—who, harking back to Edmund Burke and classical liberalism, preach a philosophy of limited government. But it is a world that is real, based not on how men and women *should* behave in theory but how they *act* in practice. It is this real world of action that has in the past frustrated both liberals and conservatives, as well as neo-liberals and neo-conservatives. Regulations have not made people intrinsically better. As the so-called Rational Expectations School of economists has pointed out, intelligent economic players soon learn how to circumvent most of the rules of the game. At best, regulation reinforces what passes at the moment for "positive" behavior. Samuel Simon, executive director of the liberal Telecommunications Research and Action Center (formerly the National Citizens Committee for Broadcasting), recently mused, "broadcasters talk about how responsive and 'good' they are. The real reason they behave decently is that they have been taught to behave that way." The "great" broadcasting system they brag about, he adds, has developed because of the law and its regulatory requirements. But by the same standard, if the overall quality of current programming isn't very good—and the existence and purpose of Simon's group show where he stands on that issue—then "the regulations" have to take their share of the rap for the situation.

On the other side, the traditional conservatives' twin doctrines of limited government and encouraged competition haven't too much meaning when it comes to some of the new media. With the new cable television technology, for example, we are currently getting more government and more monopoly—the worst of both worlds. The rights to lay cable's coaxial wires under the streets or to string them along utility poles are being franchised by local municipalities. And once franchised, cable can become, in the words of Seattle Mayor Charles Royer, "the electronic reincarnation of the company store." There may be no real competition once a cable is put down because there are not enough potential subscribers around to make it profitable. As Royer, who is also chairman of the National League of Cities' Cable Television and Telecommunications Task Force, points out, with only one wire running by homes and businesses, there will be only one source for delivery of all the programs and services cable promises. By definition, actual competition would have had to come at some earlier point, for example by creating an economic or informational context where there is relative freedom of entry. If two or more cables are not available to each subscriber, there can't be competition. The last time Royer's task force looked, such competition existed in only eight of the 6,000 American communities that have cable systems.

4

But the fact that current telecommunications reality defies current telecommunications theory should not immobilize us. There are at least five factors that are forcing changes and rewriting our ideas about regulations.

First of all, just about everyone agrees that the Communications Act of 1934, with its complicated set of broadcast regulations, needs to be changed. The present chairman of the House subcommittee on broadcasting, Rep. Timothy Wirth of Colorado, is a liberal Democrat and an opponent of deregulation. But he argues that the 1934 act, passed before the advent of over-the-air television, cable television, computers, and satellites, "is a law developed to regulate a network of country roads." On the other side of the aisle, Mark Fowler, Ronald Reagan's new FCC chairman, says that his agency "remains the last of the New Deal dinosaurs."

Second, the economic facts of life have changed in recent years. The existence of a natural monopoly in the communications field that demanded government controls, such as AT&T or the broadcasting industry in general, has diminished in the face of renewed competition which is opening the marketplace to a multiplicity of suppliers and alternative services. The current outmoded regulatory environment developed in response to a market characterized by a certain degree of homogeneity in demand—person-to-person phone calls, point-to-point voice or data communication, over-the-air radio and television broadcasting from stations within a given area to receivers in that same area of service. Today, demand is undergoing a fragmentation and specialization which has increased the number of players and produced a more competitive environment that was not foreseen when Congress determined the need to establish and protect natural monopolies in these areas. True, competition has yet to erode the dominance, for example, of AT&T in the delivery of telephone service. Even after divestiture of 22 of its local operating companies, AT&T will retain control of more than 90 percent of the long-distance market for years to come. The competition that has developed in the telephone industry and elsewhere, however, *has* lessened the need for a direct, hands-on governmental role.

Third, technology has changed the old assumptions, speeding the arrival of competition to nearly every facet of the telecommunications industry previously protected—either by accident or design—from encroachment. Technology has facilitated the entrance of competitors to AT&T in the long-distance field, the establishment of new broadcast outlets and forms that undercut earlier notions of spectrum scarcity,

5

and the creation of new markets (for example, cellular radio), and even new needs, for goods and services. What economists have called "technological exclusivity"—the basis of earlier regulation of the communications industry—no longer exists. Describing the 1934 act as "no longer adequate as a statement of national policy" today, the Senate Commerce Committee recently made the case against "unnecessary regulation" of competitive markets. Such regulation "not only represents a waste of taxpayers' dollars," but also enables firms to use the regulatory process to impede their competitors and delay new services. "The specter of regulation and appellate challenges, and the uncertainty over the commission's power to forebear from regulation," the committee adds, "have a chilling effect on investment and growth."

The FCC tried to cope with the technological changes crowding in on it by dividing the telecommunications field into two categories—traditional regulated services and "enhanced" services that could be offered in a more competitive environment. But even this relatively recent rule is crumbling as technology and innovation further dissolve the lines of demarcation between various telecommunications services. Traditional distinctions are being broken down—between written and voice telecommunication, between voice and video, between private line services and public message services, between data and voice transmission, and between computer technology and services and communications services. And the final break may come when AT&T completes implementation of its historic antitrust agreement with the government, turning loose 22 new local telephone companies and opening a new era of competition in areas from which technological giant AT&T—while fully capable of competing—has been barred since 1956.

A fourth factor is that the political climate is now right for fresh starts. The election of Reagan has signaled a historic change in the direction of government-industry relations, though again some of these changes were in the air in the Ford and Carter years, and just about everyone is in favor of some kind of telecommunications deregulation. The new view of less government, as applied to the FCC by Bernard Wunder, head of the Commerce Department's National Telecommunications and Information Administration, holds that for too long the agency and its powers "served as a sort of institutionalized cartel manager, trying to police the boundaries between the cable and broadcasting businesses, between telephone companies and cable and so forth—all supposedly for the purpose of preventing companies from 'committing competition.' " Still, while it is the Reaganauts' turn at bat—an opportunity for the conservatives earned at the polling place—not many people expect to see anything more than that tentative step we call unregulation.

In the foreseeable future, then, the Reagan team will manage a walk to first rather than a resounding extra base hit off the left field wall. In fact, a case could be made that the FCC has taken steps to ensure that it has a place in the regulatory game that is created—or forms by default—to deal with the telecommunications industry of the future. "The telecommunications industry is not ripe for total deregulation at this moment," Reagan's man Fowler recently told Congress, sounding like a New Dealing dinosaur himself. As he explained it, although the emergence of competition in telecommunications services is changing the character of the industry, "some carriers still would have the capacity to impose excessive or discriminatory rates for some services in the absence of regulation."

Finally, recent experience has shown that a moderate amount of telecommunications unregulation can take place without shaking the foundations of the Republic. Indeed, a few of us believe that the Republic can be strengthened by some of these moves. It is worth examining one such development in detail in order to appreciate some of the complex and contradictory forces currently affecting telecommunications and the new media. From this particular case we can then suggest what to expect in the entire field of media technologies, both in the near future and later. The case of current regulatory policy we want to examine concerns the limits on the First Amendment rights of broadcasters. As journalists, we are both naturally sensitive to government actions that might impede the flow of the news and information Americans receive—a concern we share with the Founding Fathers.

II. Case Study: An Argument for Getting Government Out of Broadcasting

Journalists for a long time have lived uneasily with the Communications Act of 1934 and the regulations that give second-class status to broadcasting as far as free speech rights are concerned. While journalists, and others, have been traditionally inclined to oppose governmental intrusions on the press's First Amendment rights, they nevertheless have been willing to make some trade-offs in the "public interest." Even civil libertarians have tended to go along with court-imposed restrictions on the broadcast press, such as gag orders intended to reduce pre-trial publicity; the ACLU, for instance, has publicly expressed its preference for the individual's right to a fair trial over the public's right to know.

But the Nixon years scared a lot of people and reinforced their First Amendment fervor. It was a couple of *Washington Post* newspaper reporters who helped keep the Watergate story alive; and it was, after all, the *Washington Post*'s television properties that the Nixon gang calculated would be vulnerable to counterattack, through the television licensing procedures of the FCC. The Washington Post Company's newspapers need no government license to publish; such a requirement would be contrary to the First Amendment. But television stations, the *Post*'s and everyone else's, do not enjoy such full First Amendment protections.

One of the reasons historically offered in defense of this second-class status has been the spectrum scarcity argument: Since only a limited number of voices can go out over the air, the government has a duty to regulate broadcasters to make sure they use the people's airwaves in the public interest. The classic liberal statement of this argument was made by Supreme Court Justice Felix Frankfurter, speaking for the court majority back in 1943. Referring to the 1934 Communications Act, he wrote:

> The act itself establishes that the commission's powers are not limited to the engineering and technical aspects of regulation or radio communication. Yet, we are asked to regard the commission as a kind of traffic officer, policing the wavelengths to prevent stations from inter-

fering with each other. But the act does not restrict the commission merely to supervision of the traffic. It puts upon the commission the burden of determining the composition of that traffic.

While liberals in the 1970s and 1980s have gradually moved away from Frankfurter's activist position, some conservatives have rushed in to embrace it. Recently the commentator Kevin Phillips made a New Right case for more direct governmental control of television news. To our ears, this was like being told that the sun rises in the west and sets in the east. It meant only that Phillips had made one of those sightings that conservatives are prone to make, spotting eastern "radiclib" bias in the news. As with UFO landings and reports of Big Foot, only the true believers ever make contact with this alien force; the rest of us see the normal confusions of journalism.

Phillips wants more conservative influences in the news. His argument is that the mass media, especially the television networks, are what he calls an information utility. His words are worth quoting in detail:

> Throughout the U.S. history, when "private" businesses of a certain sort—from banks and railroads to public utilities—have achieved a critical public importance, they have been subjected to increasing regulation in the national interest. Now it would seem that the information-and-opinion industry is coming under the same guns, notwithstanding self-interested assertions that it ought to be untouchable under the First Amendment. Growing public demand for media access—whether from conservatives, gays, blacks, women, or just concerned citizens—is based on growing public appreciation of the pivotal role of the media. It's just about that simple.

It's not quite so simple, of course. Phillips himself sees the irony in a situation in which conservatives, usually the defenders of property rights, are arguing for interfering with the private prerogatives of the networks. His argument also skips over the major part of the information and opinion industry—1,750 daily newspapers, 37,000 magazines, and thousands of other print outlets. Are they to be regulated by government in the national interest as well? The word "access" in these discussions usually means the right to reply—to get on the air or in print—or to have standing air time or column space. Do conservatives really believe that "the public"—not otherwise defined—should have routine access to the pages of the *New York Times*, or to Kevin Phillips's own newsletter? Changing the name of the "press" to "information utility" may make the *Washington Post* or *Time* magazine or CBS

News sound, to some ears, like Bell Telephone. But the framers of the First Amendment, and the courts through the years, treated the press as something more than an information and opinion industry; the press also was—or was supposed to be—a counterforce and check on government. In the Tornillo case, the Supreme Court overturned a Florida statute requiring newspapers to allow the right of reply. And Chief Justice Burger, a Nixon appointee no less, defended the broadcaster's right to make day-to-day decisions when he declared in a 1973 decision, "editing is what editors are for."

The principal battle in the free press versus fair press (or access press) debate now centers around federal communications law. The Fairness Doctrine, an FCC rule that has been incorporated into congressional law, says that broadcasters have to allocate a reasonable amount of time to the discussion of controversial issues and that they have to do this in a fair and balanced manner in order to allow reasonable opportunities for the expression of opposing viewpoints. The Fairness Doctrine ought not to be confused with the Equal Time Doctrine (though it often is). The Equal Time provision of the communications law applies specifically to candidates for public office and to the broadcasters' "obligations" to make air time available to candidates so that the rich or the incumbent do not monopolize all the broadcast time. If anybody were to advocate such an obligation for newspapers, the outcry would be deafening, and rightly so. Once Equal Time might have made sense when there seemed to be a limit on the number of broadcast outlets in the country. But the rule effectively kept minor candidates off the air; instead of selling time to *all*, as the rule required, most stations chose to sell to none. Now with cable TV and UHF stations opening up, there will be no scarcity of time for sale. Major candidates of course want prime time mostly. But with the larger number of outlets today, we should be willing to let the stations decide to whom, and when, to sell time.

The quick burial of Equal Time, then, should not create too much trouble for anyone, liberal or conservative. But how much harder is it for everyone—liberal or conservative—to agree on what constitutes "fair" news treatment of a given subject, or to agree on what is balance in public affairs programming? While the courts historically have been good friends of the media—print and broadcasting—they have also been sympathetic at times to the idea of an activist FCC. Because of the Fairness Doctrine, the FCC has been called upon by allegedly aggrieved parties to decide, among other things, the fairness of an NBC News report on pension plans and a CBS News documentary on the Penta-

11

gon. These topics are not ones that journalists want the political appointees of the FCC to judge.

Just about all the justifications put forth for an activist FCC to police the news are shaky and/or dangerous or out of date. Critics like Phillips still talk about the awesome concentration of network power. But the real news these days is the prospective breakup of the present broadcast system, as cable TV, two-way television, satellite networks, home recording devices (videocassettes and video discs), computer-assisted retrieval systems, and other new developments begin to fragment the old mass audience and create new special-interest viewing groups. New communications technology, as well as new private enterprise arrangements, will effectively destroy the scarcity arguments. Together they will be producing access and rights of reply without government. And Phillips still talks about further governmental incursions into broadcasters' rights when the long-term judicial trend, despite occasional backsliding, has been to lift broadcasting from second-class citizenship toward full First Amendment status.

It might be asked, if the Fairness Doctrine and other government regulation of broadcast journalism were abolished, wouldn't it mean that only the rich and the powerful could be heard on the air? We need the government to act as a referee, in the old liberal sense, don't we? Our experience as journalists tells us that the powerful, such as the oil companies and the banks, are heard so loudly and clearly now under the present system that deregulation won't make any noticeable difference. The remedy for unequal, entrenched power in America lies elsewhere in the political and economic process. It has to be addressed through electoral and market mechanisms, and not at the expense of the First Amendment.

III. Unregulation: Prospects for the Near Future

If only the debate over the future of other regulations were as clear and as easy to predict as the one about the Fairness Doctrine. The near-term prospective situation, however, will satisfy mostly those citizens able to live with a large daily ration of ambiguity. For them, some unregulation is well under way. Fowler's FCC, for example, has asked Congress for "short-form" legislation that would alter its traditional mandate. The FCC wants *specific* authority *not* to regulate areas it deems competitive—a-not-so-subtle shift away from the regulate-where-possible attitude that prevailed for 40 years.

Small steps in this direction already have been taken. Recently the commission decided it would not—after years of study and delay—select a standardized AM stereo system, choosing instead to leave that decision to the radio industry. The FCC has also sought less of a role in deciding how the airwaves are to be used and by whom; in July 1982, for example, the FCC proposed to discard its 12-year-old rule barring the three major TV networks from owning cable TV systems. The commission hopes a change in its mandate, such as that proposed in its "short-form" legislation, will help it fend off pressures to regulate in the future. This would not, however, free the FCC from its duty to oversee many regulations now in effect—or its responsibility to aid in the formation of a comprehensive policy framework for the nation as the new technologies blend with the old to create a communications and information network for which the Communications Act of 1934 is wholly inadequate.

For too long, the FCC and Congress have been lax in scanning the horizon and anticipating the changes that will be brought about by the information revolution and the technological advances that underlie it. The proliferation of new technologies and services has intensified the need for a consistent, forward-looking national policy that both encourages growth and expansion and recognizes the benefits and problems that may follow. As Judge David L. Bazelon recently noted, the central challenge facing communications decision-makers is to determine the appropriate government response to changes wrought by science. But policies can't be made if no one is looking in the right direction.

Lack of foresight is not the only impediment to carrying out such a broad objective. Politics cannot be discounted as a factor, especially in an industry whose structure and viability have been so shaped—indeed wedded to—constant regulation, and often protection, by the government. Business is quick to deride government regulation where it imposes financial costs, burdens on efficiency, constraints on profits, or limits on managerial discretion. But special interests also react by calling on friends in the Congress *to stop* the rollback or elimination of regulations when such action suddenly would place their shares of the market—often gained through exploitation of those same regulations—in jeopardy. Economist Alfred Kahn puts in bluntly: "When you scratch any businessman beneath the surface, there is a protectionist there when it comes to competing." Mark Fowler recently told the National Association of Broadcasters: "The media business is changing rapidly. . . . The attitude [that broadcasters need protection] should go the way of the Cadillac tail fin and the 10-cent Coke." But obviously a lot of people still like tail fins and cheap Coke.

The business double standard that often emerges in the exercise of the government's regulatory authority is a particularly vexing question in the telecommunications field. Nowhere was that more clearly seen than in the initial reaction to the proposed settlement of the AT&T antitrust case. Because of its pervasive position in the industry—dominant not only as a supplier of goods and services, but also as a pool of talent and innovation—any breakup of the Bell System, and particularly one as sweeping as the original settlement envisaged, would have a profound impact on the structure of the industry and its competitive environment. Thus, the rhetorical support for greater competition that had been voiced by many segments of the industry—especially those that have benefited from AT&T's technology and from its inability to compete in some fields—gave way to a chorus of protectionism once it appeared AT&T would be "unleashed," free at last to enter lucrative markets where smaller firms have prospered without its competition. This placed the courts, Congress, and the FCC in the difficult position of having to weigh a range of economic, legal, and public policy concerns and objectives, not the least of which was the reconciliation of antitrust law and regulatory motives that have committed the FCC to creating a more competitive marketplace—one marked by "free and open" access.

Under the current Communications Act, there is no reason to believe this clash of interests is going to be resolved merely by raising the level of the government's rhetorical commitment to "the marketplace." Even if the FCC makes a concerted effort to back away—to adopt a position of laissez-faire except where unusual market conditions or national needs dictate otherwise—its actions are not likely to be accepted without challenge. The result, many on both sides of this debate conclude, would be a continuing fight in the courts and in the legislative and regulatory arenas that would make the industry's future even more uncertain. As one industry observer joked to us: "I'm telling all my kids to become communications lawyers, because I don't see this wave of litigation subsiding any time soon."

Another observer, Jack Biddle, president of the Computer and Communications Industry Association (which represents some 70 manufacturers who produce $4 billion in equipment annually), contends too much time and money already have been diverted from productive enterprise—including product development and marketing—by fights over FCC regulatory decisions and such related matters as the AT&T breakup. His complaint is not isolated, but symptomatic of the way the FCC's efforts might be frustrated in the future. Still, the FCC remains committed to its new course. As Mark Fowler told the executive committee of the NAB, it makes no sense to attempt to forestall the introduction of new technology—whether it be cable, direct broadcast satellite, MDS (multi-point distribution system), or whatever the next wave of technological innovation may be—in the mistaken belief that "somehow broadcasters are entitled to special protection."

IV. The New Media Technologies: Four Shaping Developments

What then can we say about the specific development of these technologies, system by system? How will these brave new attitudes of the Reaganauts play out in practice? It is naive to think that in the years to come the growth of the telecommunications industry will be left entirely to consumers, investors, and the other actors in a theoretically pure free market. So too, we conclude, is it naive to believe telecommunications policy, heretofore the exclusive province of politicians, will be left solely in their hands. In general, regulation will continue, albeit in a different form than the structure created by the Communications Act of 1934. The new form will have a different mix and different sources of economic, technical, and social concerns; it will be a form designed to minimize impediments to growth while retaining enforcement powers in a number of areas (to be described later).

But make no mistake: The traditional priorities of regulation are gone with the 1980s wind. As the House telecommunications subcommittee, in a comprehensive review of the Communications Act and related matters, concluded: "With few exceptions, the advocates of restricted entry are gone; the debate is now between those who feel that it is government regulation which stands in the way of a fully competitive marketplace, and those who believe that a combination of deregulation and some active regulatory involvement is necessary to make the transition from essentially noncompetitive markets to fully competitive ones while continuing to protect the public."

It is likely that some of the characteristics of the new regulatory framework for the rest of the century will be the following.

1. *Less regulation of business practices, managerial decisions, and ownership.* Diversification of the marketplace and increased competition have seriously weakened the rationale used for years to regulate ownership of media properties; the AT&T settlement will remove at least a substantial part of the FCC's burden for continually reviewing the finances of its No. 1 common carrier (a task the General Accounting Office found the commission ill-equipped to do in the first place).

2. *Less centralization.* Regulatory authority will be spread among a greater number of agencies at the federal level, each with its specialized responsibilities. In addition, the continued segmenting of the marketplace will foster increased regulation at the local and state level, which will, to a degree, capture part of the terrain vacated by a retreat at the federal level.

3. *More look-the-other-way passive regulation.* Much of the emphasis will shift from direct control to a more benign oversight function, where regulation may be ordered when market conditions or other factors warrant. The FCC already has expressed a willingness to take this approach (as have other executive and regulatory agencies) on the proviso that it can always intervene if the experiment fails. This significant change in attitude will, it is hoped, create more freedom for expansion and innovation, within the confines of remaining regulations and other regulatory concerns.

4. *Exempt areas.* The word "confines" conjures up a government presence still alive and well, lurking under a broad regulatory umbrella. The new technologies, it's true, will require continued attention in some traditional areas; they will also create new regulatory problems as well, since authority will be spread among various federal agencies. Among the activities where we believe government will retain power, at least in the passive, umbrella form described above, are the following.

- antitrust and market domination
- copyright infringement and signal "pirating"
- spectrum management
- technical specifications to ensure uniformity
- fraud and consumer protection
- privacy and information security
- health and safety

The hybrid model of regulation emerging in this environment will be shaped by technological change, the abandonment of longstanding ideologies, and a more pluralistic—if not perfect—marketplace. This will shift the role of the FCC, modestly at first, away from its historical preoccupation with economics to greater technical oversight.

The FCC has become much more than Justice Frankfurter's "traffic controller" of the airwaves. It now apportions spectrum space for a proliferation of new communications technologies, from low-power TV to cellular radio; it has taken charge of the increasingly difficult tasks of controlling the use of satellites confined to a finite orbital arc and crowded into the same area of the spectrum, and delivering services that are in demand now more than ever. But the commission's own efforts to redirect its emphasis have met with spotty success, frustrated by internal delays, legal interdictions, and lingering uncertainty about how its rickety 1930s communications structure can survive in the telecommunications world of the 1980s and 1990s, and beyond.

V. Case-By-Case Analysis

With these general observations as background, it is possible to conduct a case-by-case analysis of the FCC's current and likely actions affecting broadcasting and the new media.

Broadcasting Over the Air

The FCC's most positive step toward deregulation has been—and is likely to continue to be—in its treatment of conventional radio and television broadcasting. The direction has been apparent for some time, but the trend has been reaffirmed and made clearer since the "Reagan commission" came into being. Removing layers of rules and regulations imposed on radio and television since their early days has been the central regulatory theme sounded by Fowler and others on the commission who share his free-market views. But as the commission has signaled its intention to accelerate a process begun several years ago and made easier to justify as technology has provided consumers with a multiplicity of information and entertainment sources—even in the smallest, most remote locations—it also has encountered resistance in Congress, where ultimately many of the FCC's goals will have to be implemented.

Regulation of the broadcasting industry was rooted in the scarcity argument. As we have seen, that argument no longer holds up. The real question now, says Fowler, "is how many different sources of opinion are available to the people." This issue lies at the heart of the FCC's current policy deliberations, and there is no uniformity of opinion about the answer.

Fowler's position that the scarcity rationale is obsolete is shared by his colleagues. As a result, there is definite movement away from the concept of "public trusteeship" to one of the airwaves as a competitive environment, which carries a presumption that such competition serves the public interest.

As a top priority, the commission wants to lift controls and requirements on business practices and broadcast content. The former is judged an unnecessary burden both on the commission and the industry. The latter is considered a First Amendment imperative, consistent with the commission's broader view of the desirability of allowing industry to set its own standards except where that freedom specifically undermines or clashes with the "public interest." This undertaking actually began with the large-scale deregulation of radio last year, even before Fowler succeeded Carter appointee Charles Ferris.

Deregulation of radio has been applauded by the industry and by the commission. Commissioner Anne Jones called it a major step, but added, "It turned out to be a small drop in the bucket" when compared with the task that lies ahead in dragging the remainder of the industry into the FCC's new world of unregulation. At the same time, radio deregulation angered such advocacy groups as the United Church of Christ, whose outspoken director of communications, the Rev. Everett Parker, contends that radio stations—and soon television as well—"can do anything they want," presumably ignoring the "public interest" in deciding what they air and the viewpoints they will represent.

An even more far-reaching step toward ensuring deregulation of the broadcasting industry as it exists today, and in the form it takes once the emerging technologies have taken hold, has been taken by Senate Commerce Committee Chairman Robert Packwood. In a speech to the National Association of Broadcasters last year, Packwood proposed amending the Constitution to apply First Amendment freedoms specifically to non-print media. "Two hundred years ago, this country experienced a military revolution," Packwood told the broadcasters. "Today, we are experiencing another revolution—a technological one in communications." Packwood predicted that most mass communications soon will be electronic and reminded the NAB—as if it needed such reminding—of its second-class status: "The government today says it has the right to control the electronic media—by licensing and by content."

The Packwood proposal drew praise from newspaper publishers, at least those with broadcast holdings. *New York Times* publisher Arthur Ochs Sulzberger called on his colleagues to "join with their electronic brethren to close this First Amendment gap." President Reagan, in a letter to the NAB convention last year, also moved to identify with the issue: "It is essential to extend to electronic journalism the same rights that newspapers and magazines enjoy."

The primary appeal of the Packwood amendment is that it would create a climate for the future development of the new media technologies rooted in the Reagan-Fowler commission's rhetorical commitment to the notion of unregulation. The impetus, according to Packwood himself, is the weakening of the scarcity argument for government regulation. In his speech to the broadcasters in Dallas, Packwood termed it "ironic" that the First Amendment was added to the Constitution at a time when there were only eight newspapers in the country. And Fowler adds that advances in technology make scarcity "a relative concept" that no longer provides justification for restricting the First Amendment rights of the electronic media.

Consideration of the "public interest," of course, is the usual reason for opposing deregulation. This phrase is another of those talismans of our time, like "forces of the marketplace," that are repeated by various shamans until they are meaningless. In this case deregulated radio was to lead to such not-in-the-public-interest calamities as a sudden upsurge in commercials, widespread abandonment of public affairs programming, and the dogmatic use of the airwaves. The record shows that stations have managed more restraint than the critics of deregulation predicted in their warnings of a year ago. Fowler and the FCC also believe the industry has policed itself in an admirable manner. Moreover, Richard Shiben, an architect of radio deregulation as chief of the FCC's broadcast bureau, insists market forces have proven to be adequate in protecting the public's interest in what it hears. Where numerous outlets exist, diversity in programming prospers, while in smaller markets stations generally continue to rely on more generalized formats to serve a smaller, more heterogeneous audience.

Still, deregulation in the next year or so will affect television more than radio. While the commission has publicly committed itself to following through with radio deregulation, it also has set forth an ambitious television agenda, calling for relaxation of rules on ownership, elimination of the Fairness Doctrine and Equal Time provisions, less burdensome licensing and renewal procedures, and a codification of some early deregulation moves, especially concerning the logging and community ascertainment requirements that broadcasters have had to shoulder for so long.

For its deregulation fervor, the commission has been embraced as a friend of broadcasting. However, Rep. Tim Wirth, a key figure in the Congress, continues to express concern about placing absolute faith in the marketplace, although there are few tangible pieces of evidence to support the contention that broadcasters have acted recklessly in the absence of strict standards on the airing of public issues.

For the most part, the commission is willing to treat unregulation as an experiment that it will give time to work. If the environment that emerges is unacceptable, the commissioners say they would not be reluctant, for example, to return with controls. That sentiment is evident in discussions of the commission's next moves in the area of ownership limits. There is an undercurrent of support for relaxation of the "7-7-7 rule" that limits the number of television and radio stations (AM and FM) that can be held by any single entity. However, the commissioners are divided on what should replace the current limits, and have a genuine reluctance to recommend complete deregulation. "Personally, I still think there is some need for some figure because one of the bedrock philosophies of the commission has been diversity," said Commissioner Henry Rivera, "and with concentration of owner-ship, you lose diversity." Rivera's colleague, Anne Jones, similarly talks of finding a "magic number" to replace the current limits but still prevent excessive concentration. "If there is a problem with too few voices," she said, "we can always jump back in."

The fate of the commission's rules on cross-ownership is somewhat murkier, with the urge to deregulate again at odds with deep-seated reservations about the consequences of possibly ill-advised and irre-versible action. Ms. Jones, voicing a widely held attitude at the com-mission, contends, "There is a lot to be said for letting experienced people in the market." But others, including Commissioner Abbott Washburn, still see "a need to control these extremely powerful outlets of opinion.

AM/FM Radio Expansion

The FCC's own efforts to expand broadcast services and thus weaken the hold of the powerful, a goal seemingly in tune with its philosophical bent, have been surprisingly spotty. The FCC, to take the most recent instance, was slow to evaluate and develop a position on a proposal to reduce the separation that exists between AM radio frequencies from 10 kilohertz to 9 kilohertz. The proposal would add perhaps 1,000 new stations to the airwaves—and has been greeted with skepticism related not only to concern that there already are enough stations on the air, but also to worries that today's inexpensive radio receivers might not be able to distinguish the smaller band separation, forcing consumers to purchase units with more precise tuners at vastly higher prices.

The separation issue has not been the only area of contention between the FCC and AM broadcasters, who have seen the nation switch to the smoother, cooler sound of static-free FM over the last decade. The commission's efforts to bring AM stereo into the marketplace have been a textbook case in regulation impeding the development of tech-nology, and in the disharmony/harmony that often develops between regulator and the regulated. As FM has prospered over the last few years, AM stereo has received increased attention as a key in redirecting the nation's listening habits back to the AM wavelength. At the 1982 convention of the National Association of Broadcasters, Leonard Kahn, president of one of five firms competing for rights to offer AM stereo service around the country, told the assembled broadcasters, "This is AM broadcasting's last chance." Without AM stereo, he warned, "Your industry's survival is at stake." Cullie Tarleton, chairman of the NAB's radio board, declared, "AM has got to have help"—in the form of AM stereo. "For the first time in a long time, it will give us a reason to focus a spotlight on AM and say, 'Listen to us.' "

Against that backdrop, however, the introduction of AM stereo has been delayed by regulatory and technical snarls despite repeated pledges by the FCC to get it before the public as quickly as possible. The blame lies not only with the commission, whose track record is marked with indecision and reversals, but also with the industry, which hoped to avoid risky decisions in the marketplace by having the FCC set its technical standards.

The reluctance shown by the industry to take the initiative on the issue of system selection also was rooted in legal concerns. Attorneys for the broadcasters warned that any move by the industry to select a single standard from among the competitors could land them in court under federal antitrust laws. Although some viewed the argument as specious, it gave the industry an additional reason to stay out of AM stereo as long as possible and additional ammunition for use in pressuring the FCC to retain jurisdiction over technical standards.

After years of study, the commission decided early in 1982 to sidestep a decision on which of the five competing AM stereo systems should be the industry standard. Two years before, the Carter-Ferris commission had selected a system designed by Magnavox. The reversal was rooted in the Reagan-Fowler philosophy that the marketplace, not government, should decide such issues where possible. The indecisiveness, however, kept stereo programming out of the hands of AM broadcasters—needlessly, according to some industry leaders—and kept its fate shrouded in uncertainty. From all evidence, the handwringing by the FCC, encouraged by the broadcasting and electronics industries, was unnecessary. Four months after the FCC took its hands off the AM stereo issue, the first such station went on the air. Five others were broadcasting by the end of the year. If the commission attempts to deal with these new media the same way it chose to dispose of AM stereo—by tossing them back to industry—it is worth looking at how industry behaves.

In the case of AM stereo, the industry, despite its abhorrence for government regulation, didn't want that extra burden. Broadcasters complained when the Carter FCC selected the Magnavox system on grounds of quality, but were equally unhappy when the Reagan FCC, as some put it, "abdicated its responsibility" for setting the industry standard. The problem in this case is that the five proposed systems are not compatible and could require separate receivers to listen to different stations. The task of making a single receiver compatible to all five systems could add $200 to its cost, pricing AM stereo out of much of the market.

Fowler has pointed to the AM stereo decision as a benchmark of the commission's commitment to deregulation and its faith in the marketplace. However, the commission unleashed AM stereo only after five long years of study that may have had a permanent crippling effect on its full development. The lesson may have been summed up best by one broadcasting executive who said, "The FCC has never been adept in dealing with new technologies. I'm not sure it ever will be either. . . . That is rather troubling when you consider what is yet to come" with the more complex new media technologies.

Low-Power TV

Even more troublesome have been the FCC's trials clearing the way for another, even more significant, component of the "new technologies"—low-power television. In few instances has the FCC's vision of revolutionizing the concept of access to the airwaves—so central to its mission—become so snarled in its own regulatory framework.

The attraction of LPTV is not difficult to understand. By authorizing the construction of low-power alternatives to market-dominating stations now on the air, the FCC was not only opening the door to some 8,000 new broadcast outlets, but also opening the relatively exclusive, capital-intensive world of television broadcasting to a whole new cast of characters. The FCC began investigating LPTV several years ago as a near-perfect solution to the scarcity problem that has been at the heart of many of its regulatory decisions and has governed the award of broadcast licenses for half a century. By giving an LPTV station only enough power for its signal to carry, say, five to 10 miles, the theory was that the enduring problem of spectrum clutter, the technical rationale for limiting broadcast licenses in a given locality, would be solved.

But the FCC, like other LPTV advocates, also saw an attractive secondary impact: The relatively low cost of the hardware required—a fraction of the cost of equipping a full-scale television station—would put LPTV stations in the hands of minorities, community groups, and others who over the years have fought long and hard, both before the FCC and in the courts, for greater access to conventional broadcasting facilities. Thus, the commission would be forced to weigh its traditional regulatory concerns against pressures brought to bear by social policy. But while LPTV was viewed by the commission and the voices of deregulation as a potential boon to those who have been underrepresented by the television industry, it was, at the same time, viewed as an unwelcome addition by the broadcasting industry, which maintains a formidable political lobby in Washington to protect its interests. As a result, the FCC's experience with the promising new LPTV technology is a perfect example of the problems it has encountered—some of them of its own making—in coming to grips with new technologies and their accompanying new regulatory dilemmas.

When it first proposed LPTV in 1980, the FCC did not anticipate that it would turn into what Commissioner Abbott Washburn, in a remark since echoed by his colleagues, called "a monster that's engulfed us all." The problem, in short, was that the FCC failed fully to analyze the technical ramifications of low-power television, as well as the manner in which this new technology would be made available to the public—or seized on by entrepreneurs. These failures, in turn, undercut what was to have been a major stride toward deregulation of what until now has been regarded as a scarce commodity under the "public trustee" control of the federal government.

After opening the door to LPTV, the commission was inundated with 6,500 applications, creating such an administrative mess—a literal mountain of paperwork with few guidelines to decide who would get the 3,000 to 4,000 available slots—that it temporarily froze the process. This slowed the delivery of low-power technology to the marketplace while new paper flowed in—eventually the pile grew to 12,000 applications. As prospective station owners and the broadcasting industry waited to see how the commission intended to deal with such thorny technical issues as mileage separation between stations and possible interference with existing stations in the area, Commissioner Robert E. Lee told a communications seminar: "No one could predict the response that we got. And I just don't know how we can handle it." Fowler himself later acknowledged that the commission "put the cart before the horse," explaining: "We ought to initially determine what the standards are going to be if, in fact, we are going to authorize low-power television service. Now we are faced with a terrific quagmire." Of course it was too late for orderliness.

The quagmire grew even deeper as the commission discovered that the admirable goals it once envisaged for LPTV were not coming true. Faced with the task of allocating too few stations to too many applicants, the FCC was forced to decide who should benefit from LPTV. The public interest flag was unfurled again. Should LPTV be used as a tool of social policy, with priority given to community groups, women, and minorities, or should it be viewed as merely another commercial venture, with applicants judged on the basis of whether they can deliver the service they promise?

To avoid having to make this judgment, the FCC, for the first time in its history, proposed awarding licenses by lottery. The idea was that it would give minorities and other newcomers to the process equal footing with established broadcasting interests in competing for stations.

"Establishment" broadcasters were horrified. For years, they and their lawyers had learned to perfect the cumbersome FCC licensing process. They argued that the lottery concept, a sharp departure from past practice, would provide no guarantee that licenses would be awarded to parties that could afford to provide the services they promised or that the stations that would result would meet minimum standards for technical quality of programming. Such issues, they said, could be addressed in a longer, more thorough application review.

Surprisingly enough, the groups that theoretically would have benefited from a pure lottery also objected, insisting instead on either separate proceedings or a weighted lottery—giving preference to certain groups—in considering applications from women, minorities, and noncommercial or educational institutions. Fowler argued that the lottery "is the best way to get service out fast to the people." However, Edward Hayes, an attorney for the National Association of Black-Owned Broadcasters, said his group was concerned about being outmaneuvered by "filing mills—firms that will file at every proceeding for clients who don't intend to serve the public."

The FCC can be faulted, at the very least, for initially looking at LPTV uncritically as a trouble-free panacea that would accomplish broad public policy goals. "We never thought that people would want to buy 150 of these stations and hook them up to satellites to deliver, say, country and western," recalls Washburn. "I don't have anything against country and western, but this just wasn't what was intended." Added Commissioner Henry Rivera: "The conventional wisdom is that low-power is going to solve many problems. I think the reality is that lots of licenses are several years away." In any event, he concluded, "It's certainly not going to be the great boon to minority ownership that it was originally touted as being."

Few doubt the impact low-power can have if the current regulatory obstacles are overcome and its introduction into the marketplace is unfettered. Low-power pioneer Bill McCaughan has been putting together a network of stations that he says will eventually bring television programming to every community in Alaska with a population of at least 25. "This kind of thing could work in Appalachia or the Western states like Wyoming, Montana, Idaho," he says. "Australia is asking us how they can build such a system out in the bush. It can work anywhere that's isolated by geography, culture or language. It can bridge the gap."

However, the gap will not be bridged as long as the FCC adheres to a cumbersome regulatory process that in large part has grown out of the commission's concern that LPTV will present many of the same problems that led to restrictions on ownership of conventional television stations—restrictions that in today's climate are coming under increased scrutiny. "We don't want to see this thing lead to a new round of debates about concentration and commercial vs. public use and access to the airwaves," said an FCC official. "We may have made a mistake in the way we jumped into this to begin with. But if we did, we're going to be careful before we take the lid off again." His cautionary words might have come right out of the story of Pandora's box.

Cable TV: Fables and Issues

The technology that has presented the FCC with perhaps its most difficult set of problems in 48 years—with the possible exception of television's explosive growth during the 1950s and early 1960s—has been cable. Slightly more than one-fourth of the nation's 66 million households today are wired for cable. And although the "wired nation," as envisioned during the industry's early growth years, has been slower than expected in arriving, it is widely believed that as many as 60 percent of the nation's homes may be cable-equipped by the end of the decade.

The impact of this technological evolution is beginning to be felt in the promise of diversity in programming and freedom of choice for consumers, in profits for investors, and in a new set of regulatory challenges for the FCC and Congress. Cable and its related technologies and services are the catalysts in the current transformation of the telecommunications industry. Taking note of this role in the video revolution and the new information age, Mark Fowler, in his first major speech as chairman of the FCC, told cable television executives that their industry typified why the FCC, "perhaps more than any other independent regulatory agency, will make the decisions that will affect the type of society this country will have in years to come."

Cable's proliferation, in addition to opening new horizons for users, has spawned economic competition between the "new" and "old" video technologies that has increased pressure for direct regulation, and for the FCC (as well as Congress) to play the role of referee in the new, expanded marketplace. The new is represented by entrepreneurs like Atlanta's Ted Turner who last year predicted a pot of gold at the end of the rainbow for cable systems choosing to take on the established media. The old guard rallies to exhortations like that of Jack Valenti, president of the Motion Picture Association of America. "We're heading for catastrophe in the free television business, in my judgment," Valenti says. "In the next four to seven years, free television is going to be a barren wasteland and this is going to be two nations—one divided into those who can afford pay-cable and those who can't. I think the political implications of that are dire and gloomy."

Coaxial cable—some of it capable of carrying as many as 104 channels of information, entertainment, or other material—is being laid in cities and towns across the country to link homes with "head-end" operations that range in size and sophistication from shacks adjacent to antennas that import signals from just over the horizon to huge production facilities that blend the technologies of conventional over-the-air television, microwave transmission, and satellites.

As has been the case for years, uncertainty hangs over the cable industry today. Michael Marcovsky, a cable pioneer who helped develop the two-way Qube system in Columbus, Ohio, and now serves as a consultant to others on cable ventures, says: "What is true this afternoon will be changed by next week." The state of flux in which the cable industry finds itself is heightened by regulatory unknowns—the outcomes of political and economic fights yet to be waged.

An industry trade magazine called 1981 "the year that cable would like to forget." The political climate was likened at one point to "guerrilla warfare" by Tom Wheeler, president of the National Cable Television Association. While cable has made obsolete the concept that television is a medium of scarcity—the theoretical foundation of its regulation by the government—at the same time Congress, the independent regulatory agencies, and the executive branch have not been able to come to grips with the question of how cable should be treated, even as cable systems and viewers all are on the rise. "Cable is divided on what it is," Wheeler notes. "Many people want to preserve the status quo, while others see quite a different vision."

The main argument the cable industry has had with its Washington regulators has not been with the FCC, but with Congress, which has been lobbied hard by over-the-air broadcasters and by the producers of television programming to place statutory restrictions on cable operators. This group views cable operators as the pirates of the airwaves.

As noted earlier, the fight is mostly over program exclusivity and copyright royalties. In a familiar regulatory pattern, it also has spilled over into the courts, which recently have done more than either the FCC or Congress in carving out the cable industry's rights and responsibilities. Rather than any kind of reasoned policymaking, as one telecommunications lawyer told us, "we have regulation by court order."

The FCC moved unambiguously toward deregulation of cable in 1979. That's when it adopted the conclusions of an economic analysis finding in error the basis of an earlier push to impose stringent rules on the cable industry. The idea overturned was the view that cable posed a real threat to over-the-air broadcasters.

The FCC had proposed direct regulation of cable in 1965 under pressure from broadcasters to lock the new technology into a permanent position of inferiority to the old. Cable systems then were little more than community antenna services for improving reception of distant signals. The commission said such systems "cannot be permitted to curtail the viability of existing local [TV] service or to inhibit the growth of potential service by new broadcast facilities." As the FCC saw it then, community antenna television serves the public interest "when it acts as a supplement rather than a substitute for off-the-air television service." Predictably, the FCC position was applauded by the National Association of Broadcasters as setting "a clear course . . . to preserve and advance free broadcasting, while recognizing CATV's legitimate place as a supplementary service."

The FCC's earliest rules on cable dealt with programming, limiting the number of signals introduced into a local market because they posed the potential of "fragmentation" of the local audience base supporting existing broadcast facilities. From the outset, the FCC's regulation of cable hung on to what NTIA head Bernard Wunder has called "the slender statutory thread of 'reasonable ancillarity,' " that regarded cable as subservient to broadcasting. The FCC also insisted that despite Supreme Court decisions to the contrary, cable programming was at odds with the spirit of the Copyright Act of 1909—providing additional grounds for regulation that protected existing industries (e.g., program producers).

In 1972, the commission placed constraints on the diversity of programming cable systems could offer on a pay-per-program or pay-per-channel basis—again to protect special programming offered by "free TV." When the rules were challenged in court, the FCC went on to contend that the rules were needed to prevent "undue delays" in bringing feature films to free TV. But these rules on program diversity were seriously challenged by the courts in what came to be known as the "HBO case." Next, President Ford signed legislation into law in 1976 that clarified the treatment of cable under copyright law and removed one of the commission's major justifications for its regulation of cable programming. Finally, the decisive blow to more than a decade and a half of cable regulation came in 1979, when the commission released its study of the economic and competitive costs of its rules on distant signal importation and syndicated exclusivity. The next year the rules were dropped and the path to unregulation clearly charted.

Satellites, cable penetration into the marketplace, and the concurrent development of other home video technologies all helped sweep away vestiges of the limited vision that underlay the FCC's early regulatory endeavors. But the impact of years of close regulation is undeniable. First of all, there was the slow initial growth of the cable television business—much of it attributed to "regressive Federal Communications Commission regulations, explicitly designed to hamstring the industry and to thus stymie any real or imagined adverse impact on free television," according to Bernard Wunder. In the last two years, however, this record has been replaced with a strong commitment by the FCC to the notion that cable's growth should not be restricted by regulations. The commission has taken steps to a) remove ownership restrictions, b) eliminate any lingering controls over programming and content (including its curbs on the importation of distant signals and requirements on the cablecasting of local services), and c) abolish the "syndicated exclusivity" rule granting copyright owners the right to control their products completely. At the same time, pressure has mounted in Congress to correct a perceived inequity between the relative freedom cable operators enjoy and the economic and regulatory burdens borne by broadcasters.

For the present, the FCC appears content to take a laissez-faire approach to the development of cable. The industry's growth, however, presents the commission with a number of decisions, as do regulatory pressures in Congress and at the local level. Among the issues that will continue to be of concern to the commission are the following.

Ownership

The FCC has been amenable to the elimination of restrictions on cross-ownership of over-the-air broadcasters owning cable companies and vice versa. This attitude is seen as a boon to cable's future growth, while avoiding the more fundamental question of whether any entity— for example, the American Telephone and Telegraph Co.—should be barred from the cable marketplace. The cross-ownership move deviated sharply from more than 10 years of direct regulation of cable and cleared the way for relative freedom in ownership.

The earlier restrictions had their roots in the scarcity theory, and the FCC staff concluded that with some 90 percent of the nation's households now receiving at least four television channels, the threat of misuse or domination by a single source had all but vanished. "Most markets have changed substantially since the rules were adopted and they now have a range of video outlets," an FCC staff report said. "The strong demand for new programming means there is no reason to expect cross-owned systems to restrict their offerings."

The ownership rules originally were adopted with the goal of fostering the industry's growth. But the technology of cable transformed it from a mom-and-pop enterprise into a sophisticated, capital-intensive business that was bridled, rather than aided, by limitations on who could participate in the marketplace. "The mom-and-pop days are over—thrown out the window by satellites, pay-cable, videotext and all the other advances we've seen," said a cable executive. "Growth at this point depends on attracting the big bucks."

The breakup of AT&T—a big-bucks company—presented the commission, Congress, and the courts with issues affecting the future of cable and, implicitly, the future shape of the telecommunications industry in general. Newspaper publishers eyeing the market for videotext and other "telepublishing" services looked uneasily at the prospect of going head-to-head against a leaner, meaner Ma Bell. And cable pioneers who have shouldered the cost of winning city and town franchises expressed similar anxiety at the prospect of having the telephone company—either AT&T or its offspring—using existing lines to deliver information services. The FCC is sensitive to the issue, but appears to prefer to give the marketplace time to work out its own troubles. Part of the reason the commission is reluctant to place curbs on a still amorphous telephone industry of the future is its belief that as traditional distinctions between the delivery of voice, data, and information continue to fade, a prohibition on cable ownership could at some point in the not-so-distant future place telephone companies at an unfair disadvantage to cable operators whose systems will begin looking more and more like adjuncts (or substitutes) to conventional telephone service.

In any event, the commission appears unwilling to take the lead on restricting cable ownership, even by AT&T and its offspring. "These are very sensitive issues," Fowler has said, "and I expect to look to Congress for guidance."

Copyright law and syndicated exclusivity

The biggest fight over cable has focused not on restrictive government policies, but on regulations and statutes that have afforded the industry a large measure of protection over the course of its development. Hollywood film studios and over-the-air broadcasters have watched cable take in dollars by offering programming for which, the producers and broadcasters contend, the cable operators have been undercharged.

For years, cable operators have been able to import distant signals—signals from more than 35 miles away—without bearing copyright responsibility for the shows aired. Royalty payments have been made into a single pool, then divided among copyright owners. There is widespread agreement that the system needs revision, in large part due to the proliferation of cable systems and channel capacity. And with increased concentration of ownership, Jack Valenti contends individual cable operators would not be burdened with hundreds of sets of negotiations. Rather, he says, bargaining would be in the hands of a small group of "giant" cable owners. These are not "mom and pop" operators, but companies "large enough, powerful enough and smart enough to fend for themselves in the competitive arena," in Valenti's view.

The view that will count is not the FCC's but that of the Congress, where the competing interests in the copyright debate have been fighting for protective shares of whatever legislation emerges from months of hearings. Able to read this reality as well as the next man, FCC Chairman Fowler says he believes the copyright issue is among those that should be worked out by the industry. So with millions of dollars a year in higher royalty payments—and increased fees to cable subscribers—on the line, the FCC, at least for the present, wants to leave

34

the issue in the hands of Congress, that marketplace of competing industry lobbying.

The commission has been equally adamant in its determination to eliminate another restriction on the cable industry—the "syndicated exclusivity" rule that bars cable systems from showing programs already aired in their communities. Fearing the amount stations pay to producers for programming will plummet in the face of profligate duplication, producers' lobbyist Valenti argues that such a step flies "in the face of reason, as well as the law of economics." However, unless Congress directs it to do otherwise, the FCC is prepared to allow negotiations in the marketplace to take the place of hands-on regulation.

Regulatory jurisdiction

There is little doubt that the cable industry has a friend in the Fowler commission, but the FCC—with its Reaganaut approach to unregulation—is not the only voice in the regulatory debate. Kalba Bowen Associates, a Cambridge, Mass., consulting firm that does extensive work in telecommunications, released a study last year concluding that as the FCC, and perhaps Congress, move away from regulation of the industry at the federal level, other agencies with jurisdiction over the enhanced services provided (e.g., banking), as well as state and local regulators, will rush in to fill the "regulatory vacuum." Perhaps with that on her mind, FCC Commissioner Anne Jones in 1981 cautioned the cable industry against becoming too smug with the newfound freedom it has as a result of the FCC's retreat. "The regulation of cable in 1981 looks substantially different than the rules of a decade ago would have allowed. And I think the trend is good," she said. "But there is a cost to the cable business which has not been widely discussed as far as I can tell. As federal rules are lifted, you may face some new rulemaking on the state and franchise level."

Jones's prediction is coming true. The fights in Congress over copyright royalties and ownership restrictions are well publicized, but the real day-in, day-out battle over cable's growth is underway at the local level. Cities and other local entities have tried, often with success, to expand their hold over cable by awarding exclusive franchises, exercising veto powers over cable development, and even dictating rates. The FCC's role in this is unclear. For the most part, the debate over local regulation remains a matter for the local governments themselves and for the courts. But pressure could build for the FCC or, perhaps more likely, Congress, to step in to define limits on the localities' regulatory powers or supersede them with a federal regulatory scheme.

The NCTA has decried the "ominous signs" that cities are looking to cable as a new source of needed revenue either through oppressive franchise fees or outright ownership. Thomas Wheeler acknowledges that cable systems have "decidedly local characteristics," but he insists "they are also the critical end terminals of a national system of entertainment and information." Municipal regulation must be limited, he said, "to prevent the potential subversion of national communications policy objectives for irrelevant and inappropriate reasons." Interestingly enough, the municipalities wave the same banner of "freedom" and "competition"—their freedom to act and earn revenue. Seattle Mayor Charles Royer explained his opposition to NCTA-backed proposals:

> Without any hearings or debate, a cable company will exercise sole discretion over the rates we pay for service and the rates to be paid by those who wish to use the cable to provide programming and services to the home. In simpler terms, the monopoly will decide who gets to use the only wire on the block, and the community will lose its only effective means to assure compliance with the terms of a franchise. This is a frightening prospect in a society that values the spirit of competition and freedom of speech.

Quite clearly, the municipalities want a share of the cable dollar flow; monopoly will be tolerated as long as it can be milked.

Content

Much has been said about the range of programming that can be delivered over cable, as opposed to the much more narrow latitude afforded to over-the-air broadcasting. In its infancy, cable was subjected to stringent rules on programming, including the "must carry" requirement for offering local stations programming and limits on the number of distant station materials that could be offered. Today, with cable systems capable of offering 50 or even 100 channels of programming, the need for such requirements has disappeared, and the commission has responded in an appropriate fashion by eliminating them.

Still to be resolved, though not by the FCC, are some First Amendment questions that apply uniquely to cable. As cable has proliferated, little attention has been given to the questions of how and where the First Amendment applies to the cable operators. Are they simply suppliers—deliverers—of programming or, like broadcasters, originators with certain rights and responsibilities to the public? Henry Geller, head of the NTIA under Carter, notes that the cable operator who has final control over, say, 50 to 60 channels that are delivered into 100,000 or 200,000 homes, wields a great deal of power if he so chooses. The question that arises, he says, is whether the cable operator can turn the programming spigot on and off at will—for any reason, including influencing public opinion. "That raises a very serious and sticky problem," Geller says. "Do we want one person to have the power to say, 'No, I don't want that on. I don't want Cable News Network on. I don't want this program on, I don't want this film' . . . ?" The Geller argument is that cable is a common carrier like Bell Telephone, which doesn't control what goes over its facilities.

Congress and the FCC also have taken note of complaints about pornography and other objectionable programming that is permitted over cable but would be barred from over-the-air television. Those of us who want the government as far away from control as possible hold that such problems are strictly matters for the marketplace, not for regulation. Under the First Amendment, as a general rule, the government shouldn't have any role in determining what people see and hear. In any case, there are criminal statutes that deal with obscene, indecent, or profane materials. These apply to cable television, too. For all his unregulation fervor, Fowler thinks they should, arguing that "the Justice Department should be responsible for that enforcement."

As the U.S. Supreme Court has ruled, definitions of obscenity may vary from community to community. Another policing mechanism may reside in what the market will bear. Cable subscribers—over 21, after all—should be free to contract for whatever so-called entertainment services they want. Midnight Blue, a soft-core meat rack program offered free for a time to New York City cable subscribers, may not be everyone's cup of tea; but if someone wants to program such gruel, and someone wants to pay for it, why not let these consenting adults find each other?

But what then about the children?

We think, along with Fowler, that the industry itself can take care of the problem by developing better control techniques, such as scramblers and locks that restrict program selection where necessary, in homes with children under 18. As for offended adults, no one is forcing them to watch the R-rated hijinks of the Flying Stewardesses (one of the more popular programs on the Qube system in Columbus, Ohio) or of Midnight Blue (which not so incidentally went off the air to no discernible subscriber outcry).

Privacy and security

As cable continues to grow, transforming homes into "information and entertainment centers," it will penetrate deeper into everyday life, heading us toward George Orwell's 1984.

The FCC has had little involvement in issues of security or privacy in communications; it has been content to let the laws on wiretapping, unauthorized interception of radio and telephone transmissions, and other statutes deal with the more straightforward technologies of surveillance and invasion of privacy. But the development of two-way cable has raised questions about the adequacy of existing laws and FCC regulations to deal with the threats to privacy and security that may lie ahead when the next generation of cable technology is in place.

The potential problems include not only the interception of two-way cable transmissions, but invasion of private information systems linked to a common cable system, which could violate the security of bank accounts, charge accounts, and other financial transactions that someday will come into our homes on the cable. And while one may snicker at the idea of all those good burghers in Columbus watching "R" and "X" movies, the idea that personal choices for programming made in the privacy of homes are being recorded on tape in a downtown data bank is nothing to snicker about. As Judge Bazelon asks about these centralized records: "Who will control them? The government? New entrants with no track records? Established broadcasters and common carriers?" The dangers are magnified when examined in the context of the overall telecommunications network that is likely to exist in another 10, 15, or 20 years. And although there have been few overt efforts to deal with these growing fears, the warning signs in some instances are there. "New technologies have increased the sophistication and capabilities of virtually all forms of communications," FCC Commissioner Joseph Fogarty said. "Unfortunately, these new technologies have also contributed to the development of refined tools suitable to be used for unauthorized interception and tampering with modern communications services."

It is not clear what role, if any, the FCC would have in overseeing the privacy and security implications of the new technologies. It is more likely those responsibilities will fall to other specialized agencies, such as the Federal Reserve for electronic banking or the Justice Department for personal privacy. But the commission, by virtue of its expertise in telecommunications technology, would play a lead role in any survey of the problem. "Unfortunately," Fogarty said, "aside from issuing staff papers and holding seminars, the commission has never developed an overall policy designed to ensure that the privacy and security of the network is maintained." Because the privacy and security question affects all the areas that the commission regulates—broadcast, common carrier, private radio, and cable—the commission is, as Fogarty says, "sitting on a bomb."

Subscription Television Arrives

Political and economic barriers to cable penetration, as well as major strides in the development of technology to scramble signals for transmission and unscramble them for viewing, have created STV (subscription television service), a new adjunct to cable and over-the-air broadcasting. The FCC first authorized STV in 1964. The first experimental station went on the air 13 years later, and industry leaders say that by the end of the decade 67 STV stations across the country could reach a combined market of 40–45 million homes—more than half of all the households in the country.

STV has experienced rapid growth in size and sophistication in only the last few years. In most cases, an STV operator purchases broadcast time from a UHF or VHF licensee, obtains programming from other sources, and provides it through decoders attached to each subscriber's television set. The encoding equipment is not identical from provider to provider (there is no reason for it to be) and licensees are required to air a minimum amount of unscrambled "free" programming each week.

The FCC has been heavily involved in regulating the STV industry. And as has been the case in other areas, it also has begun to back away from many of its past regulatory decisions, on grounds that they no longer are valid or desired. The commission has rules regarding technical standards for STV decoding equipment, requiring that the decoders be leased rather than sold; prohibiting STV in markets with fewer than four over-the-air stations; and mandating that STV operators show how their programming would suit community needs. The first set of rules—setting forth technical and administrative standards for decoding equipment before it is offered to the public—is viewed as a necessary and unburdensome oversight responsibility. Doubts about this specific technology are at the same time being erased by improvements both in reception and in the ability to "address" individual units—authorizing the decoders to function—at lower cost.

The remaining regulations, however, grew out of the FCC's determination in 1968 that STV—then an unknown quantity—should be held in subservience to conventional over-the-air "free" broadcasting. The commission said it viewed STV as a "beneficial supplement" to free TV at best, and as a threat at worst. All of this is about to change, however. The FCC in 1981 proposed substantial deregulation of STV, including elimination of rules on where such services could operate, rules mandating the delivery of at least 28 hours of free programming a week, rules on decoder leasing, and rules on ascertaining community needs.

The commission, in what could be a boon for the industry, said in proposing deregulation that the "protection of conventional television cannot justify economic regulation of STV." The path toward competition became clearer when the Justice Department sided with the FCC, saying that "unnecessary restrictions on STV services have stifled competition, impeded STV's efforts to gain a greater share of the video distribution market and frustrated consumers seeking video programming to suit their tastes." The brief is worth quoting for what it tells us about that department's views on unregulation:

> Some loss of conventional programming is not necessarily bad. It is in every broadcaster's interest to provide the kind of programming most desired by its customers, and thus a conversion from conventional to STV service is likely to increase net economic welfare. Operating in this manner, the marketplace is far more sensitive to customer desires than is the unyielding prohibition of the complement of four [stations] rule, which unjustifiably distorts this market process.

Predictably, the STV industry is pleased. Gone now are barriers that could not have been overcome solely through technological improvements. The industry remains worried about the rougher side of the marketplace—the use and sale of bootleg decoder systems that permit reception without paying. This "piracy" issue is under study by Congress while the industry tries to develop means of detecting illegal receivers and additional barriers to the construction of bootleg decoders. These efforts of relief appear to be well underway with the FCC confined to the cheering section.

Satellites: Open Skies, Endless Questions

Nothing has done more to change telecommunications services in recent years—in every area from voice communications to television production—than satellite transmission. In geostationary orbit 22,300 miles above the earth, satellites are delivering programming to radio and television stations that once came by land line, transmitting data for business and handling an increasing proportion of all international communications. And it all happened within just two decades, when the aptly named Early Bird satellite made transatlantic television transmission possible for the first time.

Satellite technology touches on every part of the FCC's regulatory operation, and the commission exercises wide-ranging controls over satellite use. Thus, the FCC's satellite-related activities point clearly to the changing nature of its function, both under the unregulation banner of its current chairman and as a result of changing technology.

For the FCC, this has resulted in a transition from direct regulator to traffic cop, responsible for overseeing the rapid increase in the number of communications satellites now in orbit, as well as supervising the users' push for even more space-based transmission capability to accommodate an explosion in service. Some who have pushed the commission in the direction of deregulation applaud its preference to rely on the forces of the marketplace in guiding the development of satellite technology. Others, including some members of the commission, have expressed a desire to go even further in removing the hand of regulation, but acknowledge that technical problems stand in the way of totally removing regulatory controls that may have little, if any, economic justification. On the technical side, the FCC has assumed responsibility for parceling out orbit positions and preventing "overcrowding" that would have an adverse effect on signal transmission or reception. In furthering its policy of "open entry" to telecommunications markets, the commission proposed nearly doubling the number of orbital slots available to satellites that currently operate at essentially the same frequency range, receiving signals from earth that are transmitted at 6 gigahertz and relaying them back at 4 gigahertz. Still others operate in a 14/12 gigahertz sequence. To achieve this near doubling of allottable positions, the FCC proposed cutting the spacing between satellites in half—from 4 degrees of longitude to 2 degrees. The action was a follow-up to an FCC decision to quadruple domestic satellite capability by the mid-1980s through construction of 25 additional satellites.

The rationale behind the commission's close regulation of satellite technology, like its earliest regulation of broadcasting, is the theory of scarcity. As the use of satellites has increased, so too have the demands on the FCC to manage an ever-more-cluttered spectrum. But just as the commission has re-evaluated its close regulation of over-the-air broadcasting and cable in recent years, so too has it been rethinking its legal responsibilities vis-à-vis domestic communications satellites.

The unique capabilities and requirements of the generation of satellites now in orbit weigh against a pure market-based system such as those the commission has attempted to adopt for other technologies in deciding how the medium is to be used and by whom. For example, in seeking to provide as much service to the public as possible by creating more satellite "parking places" in space, the FCC has had to weigh concerns that closer spacing would aggravate, not relieve, saturation of the airwaves by causing greater signal interference. The commission examined other ways of placing more satellites in orbit to meet burgeoning demand, but none offered immediate relief.

A staff report prepared by the Economic Division of the FCC's Common Carrier Bureau concluded that while there is no pure economic justification for regulation of domestic communications satellites, the problem of allocation of orbital space is one that, because of the technical issues involved, still must be subject to some form of regulation by the government. The allocation of "interference rights," the report said, can be worked out in the marketplace if the problem involves only two parties. However, no alternative to government regulation has been found that will protect the rights of a multiplicity of users, as well as those of the ultimate consumer.

"It is clear that if signals are transmitted simultaneously on a given frequency by several people in the same location, the signals would interfere with each other and make reception of the messages transmitted by any one person difficult, if not impossible," the report said. "The use of a piece of land simultaneously for growing wheat and as a parking lot would produce similar incompatible results."

But this "zoning" problem in space may be transitory. The picture may very well be altered by advances in technology on the transmission, reception, and switching fronts. The FCC's chief scientist, Steven Lukasik, believes "rather dramatic changes in space operations" will take place in the next decade that will lessen the regulatory pressure on the commission. What is not clear, however, is whether these improvements will have a similar impact on the international front.

The National Aeronautics and Space Administration projects a tenfold increase in the global demand for satellite circuits between 1982 and 2000. The proliferation of telecommunications services, spawned in large part by the increased use and sophistication of communications satellites, has created political as well as technical problems as the nations of the world seek to capitalize on it. These problems most likely will continue to be addressed from an approach of regional or international regulation for the foreseeable future because of the underlying geopolitical issues, while having little effect on the use of domestic communications satellites.

The FCC already has moved, albeit reluctantly, in the direction of *economic* deregulation of the domestic satellite industry, while refusing to relinquish the authority to step in with new controls if it concludes such action is warranted.

The commission considers providers of satellite services to be "common carriers" within the meaning of the 1934 Communications Act, which requires them to provide service on a first-come, first-served basis and at cost-based rates. Armed with that interpretation, the FCC moved to bar RCA Americom from completing the auctioning of seven transponders on its new Satcom TV satellite at prices ranging from $10.7 million to $14.4 million. The initial action had brought in a total of $90.1 million from seven cable television operations—$20 million more than could have been raised by leasing them under customary FCC rules. While the FCC has moved to set aside the action, it has also indicated it is not totally opposed to a more market-based leasing arrangement—taking some of the edge off what might otherwise be properly read as a significant conflict with its other calls for deregulation.

Economic deregulation is a touchy issue for the commission even under its new chairman. Commissioner Fogarty, in a speech to a seminar on satellite networking, warned of "the doom that may befall satellite users if the Federal Communications Commission continues on its perilous course of blind unregulation in the satellite field." Fogarty acknowledged that there is a danger of "over-regulation" that would have the effect of impeding the development of Direct Broadcast Satellites (DBS) and other domestic satellite services. At the same time, he views the situation as the space-age equivalent to the notion (discredited at least by a majority of Fogarty's colleagues) of scarcity of the airwaves in conventional over-the-air broadcasting.

It is because of concerns like those expressed by Fogarty that the commission's deregulation of domestic satellite service is likely to fall short of a total elimination of controls over rates—until that unknown time when technology can begin to accommodate demand and weaken the scarcity rationale. "When scarcity has come to an end, and we can be sure that rates set in the marketplace will not gouge users, I will be among the first to support deregulation," Fogarty says. "However, now is not the time."

Significantly in this regard, in the RCA Americom case the free market rhetoric came not from Fowler or his colleagues, but from the marketplace itself. RCA Americom President Andrew Inglis called the Satcom TV auction "the free marketplace in purest form" and insisted that "the company that has made the investment and taken the risk of providing facilities should reap the financial rewards of that risk." For the foreseeable future, however, with 15 satellites now in orbit and operational, the FCC will continue its tight control of spectrum space, orbital spacing, and transponder leasing. Its efforts in the latter area will be devoted in part to discouraging satellite speculators out to profit from the surge in demand for precious transponder time and in part to stop hoarding by those lucky enough to have already a part of this increasingly important medium. At the same time, the commission has proposed lifting regulations that now restrict programming available over satellites and has welcomed the day when DBS—satellite-to-home television programming—is a reality.

Direct Broadcast Service

As has been the case with other new technologies, the FCC's consideration of Direct Broadcast Satellite service has been a difficult process of assessing the technical problems and full range of impacts, as well as weighing the competing interests of DBS proponents, conventional broadcasters, and the public.

DBS poses no shortage of regulatory questions for the FCC. The problems extend into the international arena, where the implications of such issues as orbital spacing, orbital arc, effective radiated power, and beam width were to have been discussed at the Regional Administrative Radio Conference this year. Delegates to the World Administrative Radio Conference of 1977 deferred a DBS decision for the Western Hemisphere at a time when two other regions of the world (Europe-Africa and Asia) supported the concept.

The FCC is under intense pressure to limit early entrants into the DBS scramble. Because the potential for DBS is not yet known, broadcasters have asked the commission to defer action on any proposed interim operating applications or plans for long-range transmission of pay programming direct from orbiting satellites to homes via small dish antennas and "decoders" that unscramble the signal as it is received. "We've got a lot of homework to do on DBS and we're going to do it very carefully," said Lawrence Harris, chief of the FCC's broadcast bureau.

At the same time, FCC Commissioner Fogarty has warned that a misstep in the commission's authorization of DBS—too much of a risk to the marketplace—could lead to disaster:

> The prospect of DBS is exciting. It offers a significant opportunity for the larger and more effective use of radio in the public interest and

for enhancing the variety and quality of video programming sources and services available to the consumer. However, for a number of reasons, DBS is likely to be a relatively risky venture in terms of capital investment, operations and marketing development. The silver cloud of DBS may have a dark lining.

The FCC has set aside the 12-gigahertz range for DBS transmissions, but is concerned about several unresolved questions. Will the service interfere with other satellites? Would DBS be considered a broadcast service subject to such content rules as the Fairness Doctrine? Should DBS come under rules governing standard pay-TV (which the FCC has moved to eliminate)? Is the spectrum space tentatively allocated for DBS perhaps better utilized for other services?

The commission does not have as much time as it may have anticipated to resolve its doubts about DBS. Companies that view DBS as the last major technological advance of the decade in telecommunications are not waiting for the development of U.S. satellites tailored to the needs of DBS—including higher power than conventional satellites and more efficient use of the spectrum to minimize interference. Instead, they are contracting for transponder space on Canada's Anik-C satellite, scheduled for launch later this year. This carrier offers on-air dates in June 1983, while it is estimated comparable U.S. satellites will not be in orbit and operational until perhaps 1985.

Hundreds of millions of dollars—perhaps even billions on a worldwide scale—in experimentation and initial start-up costs are riding on the future of DBS. But not everyone is waiting for the technology to come of age. Across the country today, some 30,000 to 40,000 television viewers are receiving signals direct from satellites used to relay everything from Cable News Network and Home Box Office to raw "feeds" of network news reports and sporting events.

The reception dishes used to intercept these signals are far bigger and more costly than the $500, 2½-foot receivers proposed for DBS. But there is an advantage for those now willing to put down $2,500 to $12,000 for the antenna (measuring 10 to 12 feet across) and its necessary accompanying hardware (a low-noise amplifier to enhance the signal and a receiver to convert it for viewing on an ordinary television screen). The payoff to the consumer comes in selection. The dish can be directed from satellite to satellite, intercepting conventional programming, pay TV services, and virtually anything else now transmitted by satellite.

In the case of pay-TV services, there is an obvious loss of revenue from this freebooting—program suppliers who normally receive a monthly fee receive no such payment from dish owners for the right to view their product. Naturally cable companies and pay-TV services are pursuing administrative, legal, and legislative channels in the hopes of penalizing those pirates who intercept their services without paying for them.

The dish owners, however, have coalesced to fight any attempts at regulation, in part relying on wording in the Communications Act of 1934 (to make "available, so far as possible, to all people" an efficient, nationwide communications system). The pirates, too, raise their banner of marketplace freedom from the hand of government (their Jolly Roger is kept furled). Richard Brown, general counsel for the Society for Private and Commercial Earth Stations (SPACE), insists backyard earth stations are the only way several million Americans—who will never have the benefit of cable and who today have access to only one, or two, over-the-air stations—can share in the video explosion. He also insists dish owners would be willing to pay some "marketplace rate" to pay-TV providers—and indeed some have sent checks to suppliers such as HBO and Showtime, only to have them returned uncashed.

The fundamental issue, Brown argues, is the free use of the spectrum. "It is no answer that 'DBS is coming.' It is here," he told Congress. "The genie is out of the bottle. Thousands of your constituents are enjoying many channels of television for the first time. We urge you to make sure that the future that can happen, will happen."

The dispute between providers of programming and the earth station owners appears headed for the courts. It has been debated on Capitol Hill, where the dish owners succeeded in keeping an anti-piracy rider off a bill last year, but still face the judgment of Congress. For its part, the FCC does not see a significant regulatory concern in the free purchase and operation of backyard earth stations. But as with other technologies whose development it has sought to encourage without interference, the FCC has felt pressure to take some role in determining the status of the private earth stations. "We're trying to stay out of these issues," Fowler says, "though not very successfully."

These topics seem more likely to be tackled by Congress and possibly the courts than the FCC—unless the development of regional scramblers renders inoperable the pirating of signals. What technology giveth it can also taketh.

Cellular Radio

Every day, some 700 autos navigate the streets of New York City equipped with mobile telephones that represent the intersection of two trends in American life—increased mobility and increased need to communicate. But at any given moment of any given day, fewer than two dozen of these perquisites of the privileged can be in use at the same time. Mobile telephone service has been slow to catch on because of its limited availability—a condition created not by cost or regulation, per se, but by the need for centralized receiving-transmitting and switching stations with the capability to provide a wide area of service to a multiplicity of users.

Soon that may all change. The telecommunications industry—traditional players as well as several upstart competitors—is poised to jump into a new market for mobile telephone service, one likely to be more affordable and accessible to millions of additional consumers all over the country. The key to this new age in mobile communications is the development of what is known as "cellular radio"—a concept that will replace large, centrally located transmitters in urban areas with a honeycomb of wire- or radio-linked low-power transmitters. Each of these services has its own small geographic area but is capable of switching off a call from one area to the transmitter located in an adjacent area through the use of computers that detect when the signal is beginning to fade.

While cellular radio service is new, the technology is not. The FCC began studying cellular technology 13 years ago. Today it stands as one of the longest-running uncertainties in the commission's history. According to William Ginsburg, a former FCC official who joined with two of his ex-colleagues to form Cellular Communications Inc. (which hopes to share in some of the first licenses for the service): "Cellular radio has the potential for a true revolution in personal communications."

Ginsburg is not the only one with high expectations. AT&T, which developed the technology, projects that annual revenues from cellular could total $6 billion by 1990. The consulting firm of Arthur D. Little Inc. is somewhat more conservative, but even its estimate of $2 billion a year has done little to dampen enthusiasm. With 170,000 vehicles across the country now equipped with mobile telephones and another 100,000 on waiting lists (which can mean a 10-year delay in some areas), the demand for cellular is clear.

The technology is here; the problem for the FCC has been how to let it loose. The first step came in February 1982 when the commission set a June 7 deadline for the receipt of applications to serve the nation's 30 largest markets. The order was a breakthrough for cellular entrepreneurs waiting to leap in with millions of dollars in investment capital, but the way in which the commission opened the door was not without controversy. The FCC order authorized two competing systems per area reserved for the 40-megahertz range on the spectrum and went on to mandate that one system would be operated by the telephone company and the other by a non-wireline carrier.

AT&T emerged the clear winner from the commission's deliberations because of the set-aside provision and AT&T's ability to gear up quickly to deliver a service it has been perfecting for more than a decade. But the company's eventual dominance of the cellular market remains unclear, pending a final determination on how its 22 local telephone companies are to be divested under terms of its antitrust settlement with the Justice Department.

The FCC's decision in the cellular case, challenged in the U.S. Court of Appeals by smaller potential competitors for cellular licenses, seemed to be a study in contrasts. There were repeated efforts to speed the technology's entry into the marketplace, often followed by more frustrating delays. The problem was deciding who would win the coveted licenses. In finally devising a scheme to begin doing that, the commission hit on a formula that non-AT&T competitors complained could cut them out of the market even before it opens. The set-aside aspect of the licensing arrangement announced by the FCC was criticized by non-AT&T firms as being anticompetitive and contrary to the commission's oft-stated commitment to easy and open access to new telecommunications markets by as many players as possible.

The argument was not totally ignored. FCC Chairman Fowler and Commissioner Jones both opposed the set-aside scheme, insisting the public interest (and the commission's goals) would be better served by more open competition. Ironically, one rationale the majority used in reserving half of its cellular licenses for telephone companies was to avoid domination of the market by AT&T or its soon-to-be-divested local operating companies. Opening the cellular market to total competition, one commission staffer said, "could have resulted in situations where the local telephone companies—many of which are better prepared to jump into the market—would end up operating both cellular systems in a given area." Another explanation given at the time of the decision to separate the band into wireline and non-wireline systems was that the mad scramble that would take place with "wide-open competition" for licenses would lead to protracted administrative proceedings that would further delay the delivery of cellular technology to the public. Even when the marketplace wins, it loses.

The FCC's somewhat cautious approach to authorizing cellular—in which its traditional regulatory role as well as its new stance are clearly evident—has been viewed as an attempt to avoid the nightmare that arose from its well-intentioned efforts to open the market for low-power television.

The question is whether the FCC's proposal for expedited handling of cellular truly will shorten the *effective* lifetime of the regulatory preliminaries. AT&T thinks so. But some other players in the marketplace wonder whether that will still be true once the FCC is into its second or third stage of license approvals and oversight of the industry.

Home Video

The telecommunications revolution, we are told, is transforming the home television set into a home information and entertainment center. The vision conjures up a day when Americans will shop, bank, make airline reservations, and even trade stocks by a combination of television and telephone technologies. Some of the country's biggest corporations are poised to enter the market for these new hybrid services; two technologies particularly attracting attention are teletext, which will provide free, over-the-air transmission of written and graphic material, and videotex, which will provide for two-way transmission through cable or phone lines.

"The cable industry is evolving from entertainment and special-interest networks to doing something more," said Warner Amex Chairman Gustave Hauser. "Routine, bothersome things will be handled electronically within the next five years." Perhaps, but in the meantime, with more than a half-dozen major experiments underway across the country, the future of teletext and videotex could be influenced greatly by a cautious FCC and an uncertain marketplace. The problem is not market entry or development, as has been the case with such technologies as AM stereo and low-power television, but with placing the new industry on a common footing by establishing technical standards for its development.

The FCC has been asked to set such technical standards, which would clear the way for the increasingly cluttered field of emerging competitors to embark on compatible systems and services. Among those pushing for an FCC-set standard is AT&T, which plans a joint videotex project with Knight-Ridder Newspapers. Says AT&T executive Edwin Lansam: "You can't have a situation where you have a Tower of Babel."

In response, the FCC agreed to decide whether to set standards for teletext transmission, but clung to its philosophy of "unregulation where possible" in proposing an open competitive environment. Traditionally, the commission has been involved in establishing detailed standards for conventional radio transmissions. Now, with the explosion in technology that is creating many new services, the commission will be, as Mark Fowler told a House subcommittee, "considering the soundness of allowing the marketplace to develop some standardization details while the commission merely retains control over allocation of spectrum space." Fowler said he was encouraged by prospects that the manufacturers of different teletext systems will reach agreement on a common set of standards. Failure to do so, however, is likely to increase pressure on the commission to step in; otherwise introduction of these new services will be further delayed by haggling over technical standards or undermined by a lack of uniformity.

Until these uncertainties are resolved, teletext and videotex could remain technologies in search of markets. But the long-term payoff could be worth the wait. AT&T has estimated that 7 percent of all U.S. households—some 8 million households—will be served by videotex systems by 1990. Other projections range as high as 45 million homes equipped with $19 billion worth of hardware. In scanning this horizon, the FCC has preferred to deal primarily with technical issues—and only when absolutely necessary. Fowler contends that questions of who should be able to offer home information services should be decided by Congress, not the commission, thus removing the FCC from the political fray under way among newspaper publishers, AT&T, cable operators, telephone companies, and other current and future competitors in the expanded telecommunications marketplace. Given the clout of some of these special interests (the American Newspaper Publishers Association, for example, speaks out on regulatory matters for the daily newspapers back home in every congressional district), Fowler has shown good judgment. Shakespeare's Falstaff long ago established the prudent man's position for survival on the battlefield when he observed that discretion is the better part of valor.

VI. Final Observations

While we two authors have agreed on the major issues in this monograph, we wanted to offer separate concluding comments. Together, the two statements mark off our positions on the future of the new technologies and telecommunications regulation.

Norman Sandler: Economic Issues and the New Technology

Technological innovation has made possible what has become known as "the media decade" and the coming age of information. We are now developing the means to transmit hundreds of telephone calls simultaneously, send and receive data at the speed of light, bring television to remote areas that previously were out of reach of conventional broadcast signals, beam programming directly to rooftop satellite antennas, instantly poll television viewers through interactive cable systems, and unite the long-time rivals of publishing and broadcasting through videotex and teletext systems.

American industry has maintained its competitive edge in telecommunications technology and services—and that factor does not weigh lightly on the minds of those charged with determining policies that will shape the industry's future. "Telecommunications is the Number One industry in the world. We can't do anything to prohibit its growth," former Sen. Harrison Schmitt said, with perhaps a little exaggeration. While dramatic, that growth often has been haphazard and in many instances has taken place in spite of public policy rather than as a result of it. The important thing to consider about the future of the telecommunications industry is the possible inhibiting effects of federal laws, regulatory decisions, or legal rulings. The problem is not necessarily *developing* new technologies that have the potential to change the way we work, bank, shop, relax, and interrelate, but *creating an atmosphere conducive* to that development. It is a view that has emerged as a broad consensus among economists, politicians, and industry leaders— regardless, in most instances, of ideology.

The development and growth of the new technologies will depend not only on the extent (or absence) of hands-on regulation by such bodies as the Federal Communications Commission, but on the *climate* created by those actions—the psychological factor that can slow the flow of investment needed to stimulate and support research, development, and marketing. As economist Larry Darby, who served on the staffs of the FCC and the Office of Telecommunications Policy, told the Congress last year: "Uncertainty about legal and regulatory outcomes has quite probably restricted the flow of capital and the introduction of new technology in these markets."

Finally, FCC Chairman Fowler gave a Senate committee the needed assessment of what is at stake: "To prepare for its key role in the coming information age, the telecommunications industry must attract billions of dollars of investment capital to continue its rapid pace of technological change. I believe that the necessary capital and the necessary innovation are most likely to be supplied in a competitive environment, free from government regulation wherever possible." The virtual explosion in information and communications technology already has begun to challenge the assumptions that have stood for years as the technical, economic, and political guideposts of regulatory efforts. It is only recently that these efforts have begun to be viewed as potentially serious impediments to further growth of the industry. Technology itself is undergoing a rapid evolution that already may have outstripped the government's ability to keep pace in regulating it. As economists Manley Irwin and John Ela observed, telecommunications technology "has ripped apart the premises" underlying traditional theories about the exercise of regulatory power.

Although there is some evidence that this outlook is beginning to change, adherence to this traditional concept of the role of regulation—and the lack of a coherent, long-term framework for deciding how and where telecommunications will fit into the government's regulatory scheme even well into the next century—could have a damaging effect on innovation. It could also hinder the push for greater economic competition that holds the promise of lower costs and improved service for consumers and the maintenance of an atmosphere that encourages technological advancement.

Edwin Diamond: Removing the Government "Editors"

As previously noted, Chief Justice Burger once pointed out that "editors are for editing." For too long, however, the content of too many of our communications messages has been edited by any number of outsiders—regulatory bodies, the Congress, the courts. These "editors" have got to be removed from broadcasting, both the traditional and the new systems.

The broadcast establishment, like other enlightened capitalists in other industries, prefers the kind of federal regulation it can control. Most of the push to get these government editors out of broadcasting will have to come from conservatives (old style) and liberals (new style) in the Congress. I expect it will come in the middle and late 1980s as the new technologies finally undercut the old scarcity-based arrangements. Or perhaps more precisely, government will be pushed out of broadcasting when the broadcasters no longer require regulatory protection from their economic rivals—that is, when ABC, CBS, and NBC complete the switch of *their* interests to the new cable-satellite-home-communications systems, probably by the late 1980s.

When it does happen, however, the consumer may not notice very much at first, or even after a time. A regulation-free television, for example, will probably be no better—and no worse—than free print is today. Some of the new unregulated networks and station groups will pursue expediency as well as profits; others will not. Some will be community-minded and public-spirited; others will not, just as with unregulated newspapers, magazines, or newsletters.

Those of us who prize freedom and justice need not fret too much. The constraints on newspaper publishers and broadcast station owners take many forms other than government regulation. The audience (viewers/readers), the advertisers, competitors, special-interest groups, peer beliefs and pressures, the staff's own standards and tastes, all will influence the final shape of media organizations, as they should in a free society. Free broadcasting may not immediately make the world a better place. But by now we have learned to know better than to expect that happiness can be absolutely guaranteed by the marketplace any more than by government.

REFORMING
TELECOMMUNICATIONS
REGULATION

Milton Mueller

Introduction

The Communications Act of 1934 subjected the telecommunications industry to a degree of central planning unprecedented in the United States. The recent trend toward deregulation reflects an almost universal belief that this experiment in central planning was a failure. Nevertheless, all attempts at reform, even those promulgated in the name of deregulation, have left the backbone of federal regulation untouched: centralized allocation of the frequency spectrum.

The Communications Act, like the Federal Radio Act that preceded it, claims the "airwaves" as the property of "the public," forbidding private ownership and market exchanges of radio frequencies. This claim of public ownership has given rise to a centralized system of licensing, which provides the legal and technical basis for many of the FCC's other rules and regulations.

The Federal Communications Commission is the successor to the Federal Radio Commission, which was created by the 1927 Radio Act to allocate frequencies after broadcasting technology emerged in the early 1920s. The FCC inherited the FRC's frequency allocation powers in 1934, when the other telecommunications services were added to its domain. Control of the frequency spectrum plays a surprisingly large, and insufficiently appreciated, role in the FCC's regulation of telecommunications in general. The burgeoning cable industry, for example, is highly dependent upon satellite and other relay services that use the spectrum for the distribution of its programming. The FCC's regulatory authority over cable was established because of cable's potential impact upon broadcasting, which the FCC was inclined to protect due to its control of the frequency spectrum. The new competition that has blessed the once-monopolistic field of telephony does not come from companies laying new lines or creating new local exchanges, but from new uses of the frequency spectrum.[1] The different telecommunication tech-

---

[1]MCI and Sprint use microwave relays and satellites to relay a long-distance call from one local exchange to another. Satellite Business Systems (SBS), a joint project of IBM, Aetna Life and Casualty and COMSAT, uses digital satellite transmission to connect business offices around the country, in effect creating an alternative to the phone companies' local exchange. SBS offers its subscribers complete telephone, television, electronic mail, and computer hookups. Another radio alternative to the local distribution networks of the telephone companies is being contemplated by Xerox, whose XTEN (Xerox Telecommunications Network) would use the 10 Ghz region for an Electronic Message Service.

niques—satellites, co-axial cable, microwave and traditional broadcasting and telephony—are all part of an integrated network. Each technique is used in combination with the others. The FCC's absolute control over the allocation of one element of this network, the frequency spectrum, provides the foundation upon which the FCC can base its control of the rest of the industry.

During the debate over deregulation, however, the FCC's licensing power is generally taken for granted. Debate centers on which rules and regulations imposed through the licensing process should be eliminated or eased. But the fundamental issue underlying this debate is whether the frequency spectrum should continue to be treated as "public property" and centrally allocated by the FCC, or whether private, freely transferable rights in radio communication should be created and a full-fledged market system introduced.

This analysis finds that the creation of a market in radio communication through the definition of freely transferable rights is desirable for two reasons. First, it would introduce a price system into the process of frequency allocation. The incentives and signals created by market prices would lead to more efficient rationing and to conservation of this scarce economic good. Efficient rationing will become increasingly important as the new services made possible by the new technology enter the market. Equally important, definition of property rights would make open entry into radio communication services possible, thereby introducing more competition into the field. Central allocation of frequencies make all entry dependent upon the approval of the FCC. Aside from the inherent delays and costs created by such administrative review of entry, it is clear that established firms often use the FCC's power over frequency allocation to shield themselves from competition. A system of property rights will introduce a fair, orderly and swift procedure by which new competitors can enter any telecommunications market where new services are needed. This will allow the industry to adapt to changing conditions without the need for government direction. Introducing a price system and defining a flexible entry procedure will do much to bring order to the current regulatory chaos in telecommunications.

The first section of this report analyzes the relationship between property rights and deregulation, noting that deregulation is creating a *de facto* system of private property, yet one devoid of some of the most important benefits of a property system that permits free exchange.

The second section analyzes the nature of scarcity in radio communication, criticizing some of the common fallacies concerning spectrum scarcity and the electromagnetic spectrum's status as a "natural resource."

The third section is a critique of the present system of frequency allocation. It notes that the absence of a price system has created and will continue to create severe problems in spectrum management. A price system is shown to be possible only by the introduction of freely transferable rights; alternative economic techniques such as auctions and lotteries are criticized as inadequate.

The fourth section shows that a feasible system of freely transferable rights can be based on transmitter and receiver inputs. Such a system of property rights is already in use on a limited scale in the 4–6 Ghz band.

The fifth and final section notes that government regulations quickly become obsolete as the technology and economics of communications change. A system of private property or freely transferable rights, in contrast, would establish enduring rules that would protect the public's interest in justice and efficiency while allowing the industry to adapt to changing conditions.

I. Deregulation and Property Rights

The idea that private property rights can be created in radio communication is not new; it was first proposed by Leo Herzel and Ronald Coase of the University of Chicago in the 1950s.[2] Since then, discussion of the issue has been mostly confined to scholarly journals. High-level policymakers, such as the 1968 Presidential Task Force on Communications Policy, considered private property proposals, but the change seemed too radical at the time and never reached the nation's legislative agenda.

Recent trends in regulatory policy, however, make the idea seem not so radical anymore. For the past 10 years the FCC has been deregulating telecommunications at an accelerating pace, and every deregulatory initiative moves us closer to a system of private, freely transferable rights.

In September 1981 the FCC, under Chairman Mark Fowler, submitted a legislative package to the Congress that proposed sweeping changes in the Communications Act, including elimination of those portions of Section 315 that articulate the so-called Equal Time and Fairness Doctrines. But the centerpiece of the proposal was an amendment that directed the Commission to rely on "marketplace forces" as the primary element of policymaking. In effect, the Fowler package would make deregulation a permanent part of the Communications Act. Since there is no "market" without the exchange of private property, the introduction of "marketplace forces" into the allocation of radio frequencies—one of the primary factors of production in telecommunications—would require the definition of freely transferable rights.

The issue of property rights is also implicit in the controversy over the Equal Time and Fairness Doctrines. To broadcasters, Equal Time and Fairness are clear violations of the First Amendment because they give the federal government the power to intervene in their programming. To them, this power is as outrageous as a federal order to include

---

[2]Leo Herzel, "Public Interest and the Market in Color Television Regulation," *University of Chicago Law Review* 18 (1951): 802. Ronald H. Coase, "The Federal Communications Commission," *Journal of Law and Economics* 2 (October 1959).

a certain story on the front page of the *New York Times* in order to ensure a disgruntled reader of "fair representation." Those who would deny First Amendment status to the electronic media do so on the grounds that the technology of radio communication justifies a different kind of regulation than that appropriate to the print media. The federal government is allowed to put restraints and conditions upon the licensee, the Supreme Court stated in *Red Lion Broadcasting* v. *FCC*, "because of the scarcity of frequencies."[3] The argument from scarcity pervades defenses of Equal Time, as it does all defenses of federal regulation of the frequency spectrum. The argument of Rep. John Dingell (D-Mich.), chairman of the House Committee on Energy and Commerce, is typical: Repeal of Equal Time would grant broadcasters "exclusive and highly profitable use of a scarce and valuable resource in perpetuity, without any accountability."[4] Dingell's argument was seconded by Rep. Timothy Wirth (D-Colo.), chairman of the House Subcommittee on Telecommunications. He believes that "spectrum space is limited and broadcasters are privileged to operate over this most precious resource."[5]

Clearly, the broadcasters' claim of First Amendment rights will not be perceived as legitimate unless they own the channels they use, and many legislators are hesitant to grant them outright ownership of what appears to be a "limited resource." In the print media, what makes the deregulated "marketplace of ideas" legitimate is, quite literally, the *market*—a system of private property in which publications can be freely chosen from among a field of competitors. The public interest is protected not by government-enforced "fairness" rules, but by open entry— anyone can publish and distribute printed matter if they can afford to. The contention is that there is no fixed limit on the number of publications that can be produced, but that there is a natural limit on the number of stations that can operate in the frequency spectrum without interfering with each other. (This contention will be examined in detail in section two.) Frequencies must therefore be rationed by the federal government, and the free market/First Amendment paradigm does not apply.

Equal Time and Fairness are the last bulwarks of the "public ownership" concept of frequency allocation; most of the others have already crumbled, or are in the process of crumbling. The recent extension of broadcasting licenses (television licenses were extended from three to

[3]395 U.S. 390 1969.
[4]*Broadcasting*, December 14, 1981, p. 27.
[5]Ibid.

five years, and radio licenses from three to seven years), and the FCC's removal of program content regulations, ascertainment requirements, and other restrictions on the operation of a radio station, make the award of a broadcasting license more and more like the grant of a property right. Indeed, the van Deerlin bill, introduced in a previous Congress, would have made licenses permanent, and some recent bills have retained this feature. One bill now before Congress would not permit competing applications for license renewal.[6] The National Telecommunications and Information Administration (NTIA), which sets telecommunications policy for the executive branch, has argued that the FCC should allow Direct Broadcast Satellite (DBS) licensees to sublease their licenses.[7]

If licenses to use radio frequencies become permanent, if they can be leased or otherwise exchanged, if the users of those frequencies have fewer and fewer restrictions placed upon them, then we are bordering on a system of private, freely transferable rights in radio communication. Nevertheless, "public ownership" and central allocation authority remain in the Communications Act, and no coherent legislative alternative has been formulated. The result is an uneasy combination of two radically different regulatory concepts. If regulatory policy is not to oscillate arbitrarily between these two approaches, Congress will have to make a choice between them.

If Congress fails to make a decision and write it into the Communications Act, we could end up with the worst of both worlds. The FCC's regulation is double-edged; it *restricts* what licensees can do, as the industry is quick to complain, but it also *protects* licensees from competition. The FCC's frequency allocation and assignment criteria, as we shall see in greater detail later, limit the number of available chan-

---

[6]S. 1629.

[7]"NTIA believes that allowing licensees to sublease will immediately restructure private sector incentives to use spectrum more efficiently. Economic pressures will result in:

1. emergence of narrower channels, use of amplitude compandored single sideband (ACSB)-type channel reduction methods, and other innovative ways of increasing frequency efficiency;
2. implementation of effective shields, more precise thrust stabilizers, and other equipment intended to increase the number of usable orbital positions;
3. initiation of R&D into space-frequency sharing options, *e.g.* hybrid satellites and multiple ownership space platforms, leading to greater efficiency in orbit-spectrum utilization."

"Inquiry into the Development of Regulatory Policy in Regard to Direct Broadcast Satellite for the Period Following the 1983 Regional Administrative Radio Conference," *Comments of the National Telecommunications and Information Administration*, Gen. Docket #80-603.

nels and do not allow the users of these channels to subdivide and reconstitute them to make more available to new entrants. Within such a politico-economic framework, removing the restrictions on licensees without removing the protection afforded by the FCC's control of frequency allocation continues to give established firms a powerful advantage over new competitors.

An example is provided by the FCC's about-face on its own proposal to reduce the AM broadcasting channel width from 10 khz to 9 khz. The bandwidth of AM radio has been 10 khz since 1928. Radio technology has obviously advanced since then, making it feasible to reduce channel spacing to 9 khz. Many other countries already use 9 khz spacing. By reducing spacing the FCC would create room on the frequency spectrum for hundreds of new stations. Under Carter-appointed Chairman Charles Ferris, the FCC advocated just such a change and sold the entire Western region of the International Telecommunications Union on the change as well. But the new administration reversed this position. The primary impetus for keeping 10 khz, not surprisingly, came from the National Association of Broadcasters and the National Radio Broadcasters Association, organizations that represent incumbents in the industry. Had 9 khz passed, their members would have been faced with new competition for scarce advertising revenues, and the technical changes required by the shift would have imposed some costs on them. Thus, while the FCC is rapidly removing many of the restrictions on existing broadcasters, efforts to subject them to new competition are often stalled because of the system of centralized frequency allocation. The primary rationale for deregulation is that the discipline imposed by market forces serves the public interest better than direct oversight. But by failing to define freely transferable rights in radio frequencies, the government is exempting radio communication firms from one of the most important kinds of market discipline: open entry.

II. Scarcity in Radio Communication

Property rights in radio communication cannot be defined, or even discussed intelligently, unless we start with a proper understanding of what the electromagnetic spectrum is and how it is used. That the spectrum is "scarce" in some sense, and that this scarcity creates the need for some kind of rationing, is obvious to everyone. Nevertheless, while the term "scarcity" is invoked often in discussions of radio communication and telecommunications policy, there is still a great deal of confusion about what the term means when applied to radio communication.[8]

It is common to hear the spectrum referred to as a "natural resource." This is true not only of politicians like Dingell and Wirth, who believe that this "precious resource" ought to be owned by "the public," but also of many economists and engineers who tend to support private ownership and deregulation. The former chief of the FCC's Office of Plans and Policy and the FCC's current chief scientist, testifying before Congress in 1979, described the spectrum as "part of a subset of natural resources, namely those resources that do not conform to legal or geographic boundaries." Their testimony went on to compare the rationing of scarcity in the spectrum to that of other nonconforming resources such as fish, oil, and water.[9]

This characterization of the electromagnetic spectrum is fallacious and misleading. The spectrum is not a "natural resource"; it does not even exist independently of specific transmitters and receivers. The economist's and politician's treatment of the spectrum as a resource is strangely reminiscent of the 19th-century belief in the existence of an

[8]Les Brown in *Channels* magazine believes the airwaves should remain public property because broadcasters use "the air that is essential to life, the air we breathe." Of course, radio communication has nothing to do with air—it can and often does take place in outer space, where there is no atmosphere. The fact that this comment came from one of this country's most prominent writers on electronic communications underscores the need for some clarification of the economic characteristics of radio communication. "Fear of Fowler," *Channels*, January/February 1982, p. 36.

[9]Testimony of Nina W. Cornell, chief of the Office of Plans and Policy, and Stephen J. Lukasik, chief scientist, FCC, before the Senate Subcommittee on Communications on S. 611 and S. 622, to amend the Communications Act of 1934, June 18, 1979.

"ether"—an invisible, incorporeal medium through which radio waves pass. But physicists since Steinmetz and Einstein have discarded the notion of an ether; perhaps it is time policymakers caught up with them.

Electromagnetic energy consists of oscillating electric and magnetic fields which traverse space at the speed of light. The term "frequency" refers to the *rate* of oscillation and is denominated in units of cycles per second or hertz (abbreviated hz). The frequency spectrum is the scale of frequencies from 0 hz at the bottom to cosmic rays, with a frequency of 10^{25} hz, at the top. What is commonly called the *radio* frequency spectrum is simply our term for the range of frequencies suited for telecommunication, and runs from 10 khz to 300,000 Mhz.

Radio communication takes place when a transmitter and a receiver resonate on the same frequency. The phenomenon of resonance can be observed by setting two tuning forks of the same pitch near each other. Strike one, and the other will begin to vibrate. Radio communication uses this kind of energy-transfer to move information from one point in space to another, but with radio the interaction is electromagnetic and does not involve vibrations in the air. Information is encoded by the transmitter as a set of variations on the frequency oscillation pattern. These variations are called the *modulation pattern*. The frequency oscillation pattern provides a consistent frame of reference (just as a wire connection does) against which a modulation pattern can be interpreted. Thus, a receiver tuned to the same frequency as a transmitter can decode the modulation pattern and reproduce the information.

There is no "spectrum" then; there are only *transmitters* and *receivers* of electromagnetic energy. Electromagnetic energy can be generated by a variety of sources: a radio transmitter, the sun, the galaxies, neon lights, and automobile ignition systems. We measure this energy by frequency and arrange the frequencies in consecutive order on a map we call "the electromagnetic spectrum." The resulting classificatory schema makes it easier for us to understand the behavior of electromagnetic transmitters and receivers. But the spectrum, the arrangement, is our own creation. No Platonic entity or "invisible resource"[10] exists independently of a specific transmitter at a specific location.

[10]The reference is to Harvey Levin's book, *The Invisible Resource: Use and Regulation of the Radio Spectrum* (Baltimore: The Johns Hopkins Press, 1971), which, despite its unfortunate title, is the most thorough, indispensable study of the subject in existence.

Conversely, no *knowledge* of a transmission can be gained without setting up a specific receiver in a specific location.

As the frequency spectrum is neither natural nor a resource, it is not surprising that scarcity in radio communication is not caused by physical depletion, as is scarcity in natural resources. A radio station does not consume the frequency on which it broadcasts; a microwave tower does not deplete our stock of spectrum.

Instead of approaching the spectrum as a resource that is somehow "used" by transmitters, it is best to think of scarcity in radio communication in terms of what radio engineers call "electromagnetic compatibility." As the name implies, compatibility means that the operation of one radio transmitter does not interfere with the reception of other transmitters, i.e., their operation is compatible.

It is the phenomenon of interference that gives rise to scarcity in radio communication. This is not, however, as simple as it sounds, for the use of the same frequency does not necessarily result in harmful interference. In order to understand how interference creates scarcity, let us imagine that receivers $R_1 - R_n$ are all within a 100-mile radius of transmitter T_1 and all are tuned to T_1's frequency. As long as they are tuned to the same frequency, the modulation pattern that goes into T_1 will be reproduced by the amplifiers of $R_1 - R_n$. But if another transmitter, T_2, adds another modulation pattern to the same frequency, it may interfere with the ability of some of the receivers within the set $R_1 - R_n$ to reproduce T_1's signal. The level of interference is denoted by measuring the relative strength of T_1 and T_2 in a specific location. The signal/interference ratio that results, T_1/T_2, will obviously be different for every receiver because of the difference in their proximity to the two transmitters (and, possibly, their different technological characteristics).

The closer the signal/interference ratio comes to 1:1 in any receiver, the worse the interference will be. If the ratio is high enough, the receiver can ignore or filter out the weaker signal. But as the ratio approaches 1:1 the receiver's ability to differentiate between the two modulation patterns breaks down, and the signal becomes unintelligible. The point is that signals do not interfere with each other in space, they interfere with each other in *receivers*. Radio communication is scarce because of the radio receiver's limited capacity to differentiate between modulations that are not separated by frequency, or by a large enough ratio in received strength.

Although discussions of electromagnetic compatibility can easily become intimidatingly technical, the principle behind it is actually quite simple: To make their communications compatible, radio transmitters must be separated in space and frequency[11] by enough of a margin to allow receivers to differentiate between their signals. Thus, the further away T_2 is from T_1 (all other things remaining equal), the fewer of T_1's receivers it will interfere with. The lower the radiated power of T_2 in relation to T_1, the smaller its geographic range and therefore the fewer of T_1's receivers it will interfere with. If T_1 and T_2 operate on the same frequency—that is, if they oscillate at the same tempo—they must be separated in space to avoid interference. If they transmit from the same *location*, they must be separated by frequency to avoid interference. Because energy transmitted on one frequency will also generate weaker signals on frequencies that are subharmonic (or constant multiples) of that frequency, transmitters that use subharmonic and adjacent frequencies must also be separated in space to some degree to avoid interference. If T_1 and T_2 use the same frequency but the polarization is different—that is, if one oscillates vertically and the other horizontally—then the transmissions may be compatible.

To summarize, by resonating on the same frequency, electromagnetic transmitters and receivers establish a connection or channel for communication, just as if a wire had been strung from one to the other. From here on I will refer to these connections as channels. Just as the same path in space cannot be occupied by two cables at the same time, so for any given radio receiver no channel can be occupied by more than one transmitter at the same time if coherent communication is to result. Stated differently, a receiver tuned to a specific frequency can decode the modulation pattern imposed on that frequency by only one transmitter at a time. This is why radio communication is scarce. The scarce economic goods allocated by the FCC are not portions of a "natural resource" or even frequencies *per se*, but are these channels, or the opportunity to make electromagnetic connections among specific transmitters and receivers. Scarcity is allocated by separating transmitters in space and frequency by the degree necessary to achieve electromagnetic compatibility.

Although channels are scarce, it is a mistake to imply, as so many court decisions and politicians do, that the number available is rigidly fixed. The number of channels available can expand as the technology of radio communication improves and/or as the economics of telecommunication demands. The primary factor limiting the number of chan-

[11]Frequency refers to the rate or speed of oscillation. Thus, to be separated by frequency is to be dissynchronous or separated in time.

nels available is not technology or the "finite" limits of the spectrum, but the government's system of allocation. The absence of private property rights eliminates many of the incentives and opportunities to expand the number of channels and to make most efficient use of the channels that do exist.

As noted, electromagnetic compatibility is achieved by separating transmitters in space and frequency. As radio technology improves, it becomes possible to progressively reduce these separations without increasing interference, thereby making room for more and more channels within the same frequency and spatial dimensions. This can happen by reducing the *bandwidth*[12] of transmitters and receivers (9 khz is an example), and by reducing the spatial separations required of transmitters.[13] The number of channels available can also expand by developing transmitters and receivers capable of using frequencies (generally higher) that could not be used before.

The reductions in the separations required to achieve compatibility made possible by advanced technology are quite dramatic. It is technically feasible to reduce the television bandwidth, set at 6 Mhz by the FCC, by a factor of 5, 10, and even 100.[14] Since the technology required to do so is expensive, such a reduction may not be economical. The FCC's Low-Power Television (LPTV) proposal permits hundreds of new television stations due to closer spacing and low power. The most dramatic example of closer spacing, however, is provided by AM radio. In 1926 one engineering study estimated that no more than 331 standard broadcast stations would fit into the AM frequencies without harmful interference.[15,16] In 1939, the FCC declared that the AM band was

[12]Although we speak of "the" frequency used by a transmitter and receiver, no equipment uses a single hertz. All electromagnetic emissions cover a *range* of frequencies, and this range is called the bandwidth. The bandwidth of AM radio is 10 khz, that of television 6 Mhz, or 600 times the bandwidth of AM radio.

[13]Electromagnetic emissions have a finite geographic range. But this range is not finite because radio signals reach some point and stop. On the contrary, once transmitted a radio signal continues on indefinitely, its strength diminishing as its distance from the transmitter increases. The spatial boundaries are established when a) it becomes so weak that no receiver can detect it; b) it is weaker than the signals of other transmitters on the same frequency by enough of a ratio at the point of reception to be undetectable as long as the other transmitter is in operation; or c) it cannot reach a receiver because of the curvature of the earth or some other physical obstruction.

[14]Levin, p. 216.

[15]*Radio Broadcast*, vol. IX (1926), p. 475: cited in J. P. Taugher, "The Law of Radio Communication With Particular Reference to a Property Right in a Radio Wave Length," *Marquette Law Review*, April 12, 1928, p. 181.

[16]The AM allocation has been slightly expanded since 1926.

"saturated" with 764 stations.[17] Today, there are more than 4,000 stations fit into the AM frequencies. Much of this expansion was made possible by the use of directional antennas, which allow us to control a signal's propagation pattern.

The number of possible channels also expands by developing transmitters and receivers capable of using higher and higher frequencies. The range of technically usable frequencies has risen from a top limit of 1 Mhz in 1912 to over 40,000 Mhz currently.

While the number of channels available has expanded, the scope of expansion has been severely limited by the absence of a price system. If scarcity in radio were rationed by the price system rather than the government, the price of channels would rise as the demand for them increased. In this context, anyone who could find a technological means of creating new channels, or of making more efficient use of existing channels, would profit economically. In the land mobile services, for example, widespread congestion limited the profits of the manufacturers of mobile radio equipment. As long as all the mobile radio channels were full, the market for their product was limited. Thus the manufacturers—unable to obtain additional allocations from the FCC and hence forced to confront the full burden of spectrum scarcity—reduced the bandwidth of their equipment from 240 khz in 1940 to 15 khz or less today. In contrast, services such as broadcasting, in which the dimensions of channels are carefully controlled by the FCC, have not experienced any bandwidth reductions.

Rising prices, then, stimulate expansion or more intensive use of a good. In this respect, radio channels are no different from any other scarce good. As the price of land in congested urban areas rises, for example, its use becomes more intensive. Technology is used to create more of the good, as when a skyscraper creates hundreds of office units where before there were only a few.

A proper understanding of scarcity in radio communication also makes it clear that there is no significant difference between radio communication and print in this respect. Clearly, the physical resources that go into the production of the print media—newsprint, presses, distribution trucks, and so on—are scarce, and hence their owners charge a price for them. It is often asserted that print differs from the electronic media in that virtually anyone can prepare a written or printed message. While true, this ignores the fact that once a message is printed it must be physically transported from the publisher to the

[17]Levin, p. 220.

readers to have any effect. Thus, printed communications are as dependent upon channels of *transportation* as radio communications are dependent upon electromagnetic channels. Both kinds of channels are scarce, and for exactly the same reasons: It is costly to build a road or fly an airplane from point A to point B and costly to send a bundle of newspapers on that road or plane, just as it is costly to establish a radio channel and transmit information electronically. Scarcity as reflected in the *price* of transportation excludes some people from access to printed communications just as scarcity in radio channels excludes some transmitters in favor of others. Indeed, the rising price of energy and other raw materials makes the distribution of printed matter more expensive and less accessible to the public than telecommunication, the price of which is rapidly falling. Increases in first- and second-class postage rates, for example, have threatened the economic viability of many marginal publications.

Nevertheless, the argument runs, if no commercial alternatives are available for printed matter, an individual can always hand out mimeographed pamphlets on a street corner, whereas there is (allegedly) no room for backyard broadcasters. This argument does not stand up to analysis, and not only for the obvious reason that mimeographed copies must be bought. Radio communication, in the form of broadcasting or satellite, covers a much larger geographic region and reaches a larger audience than an individual handing out pamphlets on a street corner or mailing copies to his friends. To cover the same-sized audience as a broadcaster, our pamphleteer would have to command enough resources to print up millions of copies and pay hundreds of thousands of dollars in postage or other shipping charges. This kind of distribution is not accessible to "anyone" any more than mass broadcasting is. To make an honest comparison of telecommunication and print in this respect, we must base the comparison on each medium's coverage of a geographic region of equal size. Once we limit our comparison to these cases, we find that distribution of a message by electronic means is easier and less expensive than print. A bullhorn or any other PA system will cover a street corner as well as the distribution of pamphlets, and an individual can distribute information among his friends by means of CB radio, ham radio, or telephone as easily as he can by sending them a manuscript in the mail.

There is nothing mysterious or exceptional about the nature of scarcity in radio communication. The idea that the scarcity of channels somehow justifies federal intervention into radio, while the scarcity of newsprint and transportation channels does not justify federal regulation of print, is simply an atavism. It may have been understandable at the dawn of radio in the 1920s, when our regulatory system was formed, but there is no excuse for it now—especially now that newspapers are relying on telecommunication to an increasing degree.[18]

Despite the increasing price of transportation and the falling price of telecommunication, there are thousands of nationally distributed publications reflecting a great range of opinion and subject matter. Information that is telecommunicated cannot approach its diversity and scope. Incredibly, this fact is often cited by critics of deregulation as evidence of the need for continued regulation.[19] Apparently they are unaware of the fact that the print media are not regulated by the FCC but by the price system, that is, by the exchange of private property in the market. The flexibility and diversity in print that has arisen from this arrangement is not an argument for regulating the electronic media differently; it is a powerful demonstration that free market and First Amendment concepts should be extended to the telecommunications industry immediately.

[18]The *Washington Post* uses the 12.2 to 12.7 Ghz band to transmit its information in digitized form to a remote printer. The *Wall Street Journal* uses telecommunication to distribute its nationwide editions to local printers. And of course the wire services have always relied on telecommunication.

[19]Rep. Dingell: "[E]ven if optimistic projections for the growth of cable, MDS, STV and DBS are accurate, we will continue to operate in a climate of scarcity for some time. Hence, we will need the protection afforded by the equal time and fairness provisions against abuse of that scarcity." *Channels*, December/January 1981/1982, p. 7.

III. Centralized Frequency Allocation: A Critique

The debate over property in the "airwaves" is frequently muddied by dichotomizing the alternatives of "public" and "private" property. Public property is a euphonious term that implies that all of us acquire control over the airwaves, while private property sounds selfish. But public property is a meaningless term. Ownership means that the owner has some control over the good in question. If a resource is scarce it cannot be controlled by everyone equally, no matter what form of regulation we adopt.

If by public ownership we denote something like the national parks, which any member of the public may enter and enjoy at his own convenience, then the airwaves were owned by "the public" from 1920 to 1926, when anyone who applied to the Department of Commerce could acquire a broadcasting license. This kind of public ownership quickly led to chaos, as more than 700 stations crowded into the two frequencies set aside for that purpose by the Department of Commerce.[20] In 1927 Congress responded to this crisis by passing the Radio Act, and the Federal Radio Commission it created promptly threw about 15 percent of these stations off the air and ceased to issue licenses for several years. The airwaves have never been "public property" since. The chaos of 1926 demonstrated the need for—indeed, the *inevitability* of—some kind of rationing, some way of excluding some members of the public from the use of radio frequencies.

The choice we face, then, is not between a system of public ownership and a system of private ownership, but a choice between two different kinds of private ownership. In one case, a monopoly over a channel is granted to a private licensee by the federal government, which retains some—but increasingly less—power over how the channel is used. The federal government rations the scarce good by defining a property structure through administrative procedures that we will explore below. In a "private" property system, the difference is that license rights

---

[20]For a detailed account of the chaos of the airwaves, see Jora R. Minasian, "The Political Economy of Broadcasting in the 1920s," *Journal of Law and Economics* 12 (October 1969).

could not be withdrawn by the FCC, trading of these rights would be allowed, and hence a price system would replace administrative rationing.

The Present System

The FCC controls the use of the frequency spectrum by specifying what one can do with a radio transmitter. The rights granted by the FCC license specify antenna height and location, power level, operating hours, and other technical standards. These specifications, which we will refer to as the *inputs* of the transmitter throughout this report, determine the range of a radio signal insofar as its range can be socially controlled. The FCC arrives at these input specifications by means of two processes: *allocation* and *assignment*.

The allocation[21] process sets aside a certain block of consecutive frequencies for the use of a specific communications service.[22] The block allocated to AM broadcasting, for example, extends from 535 khz to 1605 khz. With the exception of AM broadcasting, which developed before any regulatory apparatus existed, allocation precedes commercial development of a service. Sometimes different services share the same block, but for the most part allocation restricts each service to a separate range of frequencies. An official record of the FCC's division of the spectrum is contained in the "Table of Frequency Allocations," Section 2.106 of the FCC Rules and Regulations.

Once a block of frequencies has been allocated to a particular service, the assignment process determines the bandwidth and geographic range of the particular channels within the block. The FCC sets standards governing the bandwidth of the service and the distance each transmitter must be separated from the other transmitters on the same, adjacent, and harmonic frequencies. The co-channel separation for UHF television, for example, is about 150–170 miles. In broadcasting, assignment procedures attempt to define geographic zones, called "signal contours" or "service areas," within which a station is protected from interference.

Assignment procedures vary by service. FM broadcasting and VHF and UHF television are assigned according to a pre-engineered *assign-*

---

[21]Allocation in this specific sense should not be confused with the generic meaning of the term, which applies to any method of rationing a scarce good.

[22]The FCC's allocations are made within constraints set at the World Administrative Radio Conferences of the International Telecommunications Union (ITU). Thus, allocation is coordinated at the international level as well as the national level.

ment table. These tables prearrange all the input relationships among transmitters, defining a fixed number of channels. These channels are then handed out to private users in the licensing process. The television assignment table was adopted in 1952, the FM radio table in 1963. Over 2,000 channels were made available by the TV assignment tables. However, most of them were not located in markets capable of sustaining a station, while the few that were located in desirable urban markets have long been occupied. Thus, as of 1980 there were just over 1,000 operating UHF and VHF television stations.

AM radio and the newer microwave-satellite services, in contrast, do not use assignment tables. They are assigned on more of an *ad hoc* basis. New stations are worked in in a way that will avoid harmful interference with existing stations. Land mobile radio services are allocated channels, and the coordinator of a mobile service is allowed to fit as many individual users into a channel as he can. In effect, there are no assignment procedures for most mobile services.

A distinction is in order here. "Allocation" and "assignment" are the terms the *FCC* uses to describe what it is doing. Because its regulatory program is founded on the notion that the spectrum is a "resource," allocation and assignment are the terms it uses to describe the process by which portions of that "resource" are created and handed out to private users. It would be simpler, however, and more in accord with the analysis presented in section two, to say that what the FCC really does is decide what technical standards the manufacturers of radio equipment must use and, once the equipment is manufactured, how far the transmitters must be placed from each other. The FCC's technical standards include, but are not limited to, bandwidth specifications. This distinction is important because it underscores the degree to which the FCC's role in "frequency allocation" gives it a major role in the design and production of radio equipment. The Radio Technical Planning Board established by the FCC in 1943, for example, set the technical standards for television that prevail to this day.

Allocation and assignment are based on engineering and legal criteria. But because they must be used to ration channels, a scarce good, the engineers, lawyers, and politicians who make the decisions must become full-time economists as well. By limiting the number of frequencies available to a service and by setting the bandwidth and geographic separations required of transmitters, the FCC sets an upper limit on the number of competitors who can enter a given radio market. The VHF assignment table, for example, starts from an allocated base of 12 channels. Once the transmitter separations are taken into account the number of VHF channels in a given geographic region is reduced from a range extending from 7 in a few cases to as little as 3 in others. Likewise, if the FCC allocates 500 Mhz to DBS and limits the number of orbital slots DBS satellites can occupy, it limits the number of competitors who can enter that market.

By segregating services into distinct blocks of frequencies, the FCC puts itself in a position where it must judge the relative economic value to society of different services. If it makes too many frequencies available to land mobile services, there may not be enough for relay purposes or broadcasting. Allocations made to television had to be taken away from FM radio, and the frequency allocation planned for DBS will have to be taken away from established microwave services.

While the decisions made by the FCC are thus economic in nature, no price for channels or frequencies (or for closer geographic spacing) is ever charged. A small license fee is charged to cover the FCC's license processing costs, but this fee in no way purports to represent the actual value of the channel. All allocation and assignment decisions are made by administrative procedures and are likely to involve public hearings in accord with the Communications Act and the Administrative Procedures Act. "Trafficking" or trading of licenses or channels is forbidden. Changes in ownership require the approval of the FCC. The license cannot be subleased or its input specifications altered in any way in exchange for money. Unlike virtually every other commodity in the economy, then, radio channels do not become more expensive as the demand for them increases.

In sum, allocation and assignment define the *property structure* of radio communication—they create channels and assign control of them to private users.[23] The FCC's power to define the dimensions of radio channels, their arrangement on the spectrum, and the kind of signal they can carry is equivalent to the power the federal government would have over the rest of society if it defined the size and shape of all land parcels and approved all land transactions. Because the property structure is defined exclusively by government, radio communication provides one of the purest examples of economic planning in the U.S.

Serious questions can be raised concerning whether that much power *ought* to be centralized in one agency of the federal government. This

[23]See Cornell and Lukasik, p. 15: "Legally, [FCC licenses] are not recognized as property rights, that is, as rights that could be bought and sold, but in reality they could and sometimes do serve the same purpose. Essentially, the Commission conveys a form of property right when it issues a radio license in the broadcast, common carrier and some private services, but one specified in terms of input and not output."

critique will focus on the narrower question of whether a centralized agency *can* exercise such power rationally. By abolishing private property and exchange in radio communication, the FCC also abolishes the possibility of attaching prices to channels. Without market prices, there is no way to continuously bring the dimensions and distribution of channels in line with the continuous changes in the economics and technology of telecommunication. Assuming the best intentions on the part of the regulators, the rigidity of central allocation limits the availability of telecommunication services and impedes the industry's adaptation to changing conditions.

The Need for Prices

One of the most important discoveries of modern economic science is the role of prices in facilitating rational allocation of scarce goods in a complex economic system. Prices are a medium for the transmission and reception of information no less than a satellite link or a cable network. Instead of electrical on-off patterns, the price system employs numerical variations in a common medium of exchange—money. The information conveyed by money prices is information about the supply and demand for scarce goods. Prices provide objective information about the supply and demand for a commodity because they represent an *exchange ratio,* the quantity of money required to induce the owner of a good to give it up under certain conditions. Attaching a numerical value in this way to commodities and services that would otherwise be incomparable allows individual decision-makers to directly compare the cost of alternative factors of production. In sum, money prices convey knowledge about the value of goods.

The deregulation of telecommunications began in the late '60s.[24] While technological change helped stimulate this trend, another important factor was a body of literature that has become known as "spectrum economics."[25] This literature applied economic analysis to radio com-

[24]The best review of the deregulation of communications is provided by Douglas W. Webbink, "The Recent Deregulatory Movement at the FCC," in *Telecommunications in the U.S.: Trends and Policies,* Leonard Lewin, ed. (Dedham, Mass.: Artech House, 1981).

[25]The "spectrum economics" literature includes: Herzel, p. 802; Coase; and Harvey Levin, "Federal Control of Entry in the Broadcast Industry," *Journal of Law and Economics* 5 (1962): 9–67. Herzel, Coase, and Levin are all academic economists associated with the "Chicago school."

By 1968 engineers and spectrum managers, faced with growing demand for radio communication services, found the paradigm of spectrum economics useful. This is reflected in the literature of the time: Joint Technical Advisory Committee, *Spectrum Engineering: The Key to Progress* (New York: Institute of Electrical and Electronics Engineers, 1968);

munication for the first time, treating the electromagnetic spectrum as a scarce good. It accumulated an overwhelming body of evidence showing that government decision-makers have no way of knowing the value of scarce frequencies, and hence administrative rationing is chaotic and inefficient. Indeed, the most damning evidence of the problems with central allocation comes from within the spectrum management bureaucracy.[26]

While the literature of spectrum economics is recent, it is really only an extension, within the microcosm of the frequency spectrum, of the debate over economic planning that began in the 1920s. In the early decades of the 20th century, about the same time that the commercial use of the frequency spectrum began, economic planning was a new and, some thought, "scientific" idea. The first systematic critique of planning was advanced by Ludwig von Mises, the Austrian economist.[27] Mises argued that rational economic calculation is impossible without prices that arise from actual market exchanges. Without some means of comparing physically non-comparable factors of production and of expressing those comparisons in precise units, Mises claimed,

---

President's Task Force on Communications Policy, *Final Report* (Washington, D.C.: U.S. Government Printing Office, 1968); DeVany, et al., "A Property System Approach to the Electromagnetic Spectrum," *Stanford Law Review* 21 (1969): 1499–1561; Levin, *The Invisible Resource.*

By the late 1970s the perspective of spectrum economics had permeated the spectrum management bureaucracies of the FCC and the Office of Telecommunications Policy (later to become the NTIA) in the executive branch. Webbink, in "The Recent Deregulatory Movement at the FCC," p. 3, notes "the number of experienced economists that have been hired throughout the Commission" and "the increasing emphasis on economic analysis in Commission proposals and decisions."

Studies and internal reports reflecting the economic perspective proliferated: John O. Robinson, "An Investigation of Economic Factors in FCC Spectrum Management," FCC Spectrum Allocations Staff, Office of the Chief Engineer, Report #SAS 76-01, August 1976; Webbink, "The Value of the Frequency Spectrum Allocated to Specific Uses," IEEE Transactions on Electromagnetic Compatibility, vol. EMC-19 (August 1977); Donald R. Ewing, "Controlled Markets for Spectrum Management," NTIA, 1979; Dunn Agnew and Gould Stibolt, "Economic Techniques for Spectrum Management: Final Report," Mathtech, Inc., December 20, 1979; Webbink, "Frequency Spectrum Deregulation Alternatives," FCC Office of Plans and Policy, October 1980.

[26]In addition to the sources cited above, there are many unpublished papers, texts of speeches, and manuscripts circulated within the FCC and NTIA. See, for example, "Remarks by Dale N. Hatfield, Associate Administrator of the NTIA before the SIRSA 1980 Annual Membership Meeting, Panel on Policy Developments in Mobile Radio," October 10, 1980. (copy obtained from NTIA)

[27]Ludwig von Mises, *The Theory of Money and Credit* (1922; reprint ed., Indianapolis: Liberty Classics, 1981). See also Mises, *Socialism* (reprint ed., Indianapolis: Liberty Classics, 1981).

a central planner cannot know how to put productive resources to their most efficient use. Far from bringing order to the "chaos" of the free market, he predicted central planning would itself lead to chaos.

One of Mises's followers, Nobel laureate F.A. Hayek, noted in 1945 that the price system acted quite literally like a telecommunications medium, registering information about the relative scarcity of goods and distributing this information throughout the society.[28] Hayek also called attention to the fact that the information needed to plan an industry is never concentrated in a single place, but is widely dispersed as "bits of incomplete and frequently contradictory knowledge which separate individuals possess." Hayek concluded that this "Planner's Dilemma" cannot be solved by "first communicating all this knowledge to a central board which, after integrating all knowledge, issues its orders. We must solve it by some form of decentralization."[29]

What is interesting about the history of the idea of planning is that Mises and Hayek in effect made a *prediction* about the impossibility of planning, and this prediction was tested by subsequent events. Mises's warning was ignored by political decision-makers; central planners were given control over the allocation of scarce resources. In the U.S. telecommunications industry in particular, their power to define the dimensions of channels gave the planners near-absolute control over the property structure of radio communication. The frequency spectrum is an ideal test case because it was exempted from market forces so completely.

As predicted, no price system could develop for channels without market exchanges. The absence of a price system, as the spectrum economics literature attests, did indeed make economic calculation impossible. The ultimate result was chaos in frequency allocation. The federal government intervened in 1927 to bring order to the "chaos of the airwaves" that had developed in the absence of property rights. It

---

[28]"We must look at the price system as . . . a mechanism for communicating information if we want to understand its real function. . . . The most significant fact about this system is the economy of knowledge with which it operates, or how little the individual participants need to know in order to take the right action. In abbreviated form, by a kind of symbol, only the most essential information is passed on, and this is passed on only to those concerned. It is more than a metaphor to describe the price system as a kind of machinery for registering change, or a system of telecommunications which enables individual producers to watch merely the movement of a few pointers, as an engineer might watch the hands of a few dials, in order to adjust their activities to changes of which they may never know more than their reflection in the price movement." F.A. Hayek, "The Use of Knowledge in Society," *American Economic Review* 35 (September 1945).

[29]Ibid.

succeeded in keeping chaos off the airwaves by establishing technical standards and by rigidly limiting the number of transmitters on the air. But the locus of the chaos merely shifted, into the corridors of the FCC itself.

The FCC has been unable to keep up with the pace of change in radio communication since the end of World War II. It took the FCC nearly 10 years to finalize allocation and assignment criteria for television. For four of those years, it had to impose a "freeze" on the licensing of stations. It was almost 30 years before the FCC was able to change those specifications with the LPTV proposal. It took the FCC three years to settle a dispute between FM radio and VHF television over the same frequencies, and it took 10 years to reallocate some frequencies from UHF television to mobile radio. Access to channels is thus constricted by a bureaucracy which frequently needs 10 years to make a major decision, and the result is a backlog of applicants that can only be described as chaotic. Eight to 10 applicants frequently apply for desirable television channels in urban markets, and the astronomical trading price of these stations attests to the value of the channel. Over 400 applications for the Multi-Point Distribution Service (MDS) have been received since 1978, over 100 of them mutually exclusive.

The degree to which the market is stifled by the FCC is indicated by what happens when it lifts some of its regulations and opens the door to newcomers, as it did with the LPTV proposal. The FCC was flooded with 6,000 applications. True to form, the FCC reacted by slapping a freeze on further applications because it is unable to process them all. FCC research suggests that if the freeze is lifted, another 8,000 applications might be received. Government and industry prognosticators estimate that the licensing logjam will not be broken up until mid-1983.[30] In all of these cases, the FCC must make what government spokesmen admit are mostly arbitrary selections among the competing applicants,[31] yet resolution of the competing claims can take several

[30]*Broadcasting*, November 16, 1981, p. 64.

[31]"The Commission too often is faced with selecting from among equally well-qualified applicants. In effect, they must distinguish the indistinguishable and decide the undecidable. The results are incredible delays and excessive costs that serve mostly to postpone or deny service to the public, raise prices to users, consume FCC and court resources, preclude small firms from going into business, and simply enrich a legion of communications attorneys." Dale Hatfield, Associate Administrator of the NTIA, in a speech before the Associated Public Safety Communications Officers 48th Annual National Conference, August 18, 1980.

years and cost hundreds of thousands of dollars.[32] In some cases, the cost of obtaining an assignment exceeds the value of the license.

There is also a tremendous maldistribution of channels arising from central allocation. For the most part, allocation and assignment criteria are uniform throughout the country when it is obvious that the services which can be sustained differ radically from region to region. Frequencies set aside for forestry use, for example, were only recently made available for taxicab services in New York City. The possibilities for networking a signal created by microwave relays, satellites, cable, and the phone system mean that any broadcast station in any part of the country can aspire to a national audience. Conversely, any audience in any region, no matter how remote, could receive a range of channel choices as wide as the nationwide market could sustain. Until recently, the FCC's policy of localism, enforced through its allocation and assignment criteria, has confined broadcasting to a small area around the community where the transmitter is located.[33] Thus, instead of 50–100 national radio and television channels, the public has been limited to three or four local television stations, 10–15 AM and FM stations, and only three national networks.[34] It is worth noting in this respect that the most promising proposals for low-power television are precisely those that rely on satellite and translator networks to distribute a signal produced in a larger market throughout the country. Without these

[32]The cost of a hearing for a contested 10-mile microwave link was estimated by the Mathtech study to be $342,307 when the applicant loses the dispute and $214,907 when the application is won. Estimates assume that the dispute lasts one year; many last longer than a year. See Table 10, "Cost of MX Hearings Delay for a New 10-Mile FS Link," in "FCC Inquiry into the Development of Regulatory Policy in Regard to Direct Broadcast Satellite for the Period Following the 1983 Regional Administrative Radio Conference," *Comments of the National Telecommunications and Information Administration,* Gen. Docket No. 80-603, p. 110.

[33]See Ida Walters, "Freedom for Communications," in *Instead of Regulation,* Robert Poole, ed. (Lexington, Mass.: D.C. Heath and Co., 1982).

[34]The FCC's Network Inquiry demonstrated how the rigidity of allocation and assignment criteria limited competition to three major networks. A strong national network would require access to the top 50 markets, where most of the viewers, and therefore advertising revenues, were located. Under the FCC's 1952 TV allocation and assignment scheme, only seven of the top 50 markets received four or more VHF assignments. Twenty received 3 VHF assignments, 16 received 2, and 2 markets received only 1. "As a consequence of this scheme, one network could reach 45 of the top fifty markets with VHF stations and the second could reach 43, while a third network could reach 27 and a fourth would have access to VHF stations in only 7 of the top fifty markets." The same FCC report documented how the DuMont network crumbled in the '50s as a consequence. FCC Network Inquiry Special Staff, "The Historical Evolution of the Commercial Network Broadcast System," October 1979, pp. 77–79.

programming networks, few of the proposed local stations will be viable.[35]

The role of prices in facilitating value comparisons would suggest that without them, the FCC would have serious problems making an allocation when different radio communication services desire the same frequencies. If there is congestion *within* an allocation, the FCC can adopt stricter technical standards to squeeze in more channels. But when alternative, mutually exclusive uses for frequencies are proposed, we would expect the FCC to be totally at sea. This is exactly what we do find. Conflicts over service allocations have led to extreme confusion and delay.

To date, the FCC has resolved only two major conflicts between commercial services. The first involved FM radio and television. FM radio was developed prior to World War II as an improvement over AM broadcasting. In 1940 it was granted the VHF frequencies 40–52 Mhz. By 1943 there were more applications than available channels in New York and New England; by 1945, 55 stations and an estimated 400,000 receivers existed.[36] At the end of the war, both FM and the new television industry were ready for more frequencies, and the VHF range (roughly, 20–120 Mhz) was believed best suited to both. A long political battle ensued. The crossfire of testimony, technical reports, congressional hearings, and court challenges lasted three years, from 1944 to 1947. In the end, the FCC moved FM from 40–52 Mhz to its present slot, 88–108 Mhz, and awarded TV the frequencies 44–88 Mhz (Channels 1–6) and 174–216 Mhz (Channels 7–13). The FM industry was wiped out by this decision and did not recover until the late '60s. Consumer investment in receivers was rendered obsolete, and established stations were forced out of business.

The only other time the FCC was faced with a conflict between two services, its performance was even worse. In 1956 the FCC initiated allocation proceedings; one proceeding, Docket 11997, documented the need for more frequencies for land-mobile radio (LMR). Land-mobile advocates requested unused channels allocated to UHF television, but the docket was terminated without making a reallocation. A land-mobile advisory committee was created, though. Years later, after mounting pressure, the FCC authorized the use of UHF channels 70–83 in the 50 largest markets, and 115 Mhz was given to LMR from the

[35]*Broadcasting*, February 23, 1981, pp. 58–62, carries a list of the proposed LPTV networks.

[36]Christopher Sterling and John Kittros, *Stay Tuned: A Concise History of American Broadcasting* (Belmont, Calif.: Wadsworth Publishing Co., 1977).

government and broadcast studio-transmitter links. Reallocation of 200 Mhz thus took about 10 years. Some of the frequencies reallocated are still not being used,[37] and LMR services are still highly congested.[38]

The LMR-UHF reallocation graphically demonstrates how difficult it is to get a centralized, administered system to make the kinds of adjustments that a price system would induce automatically. Although the FCC has no idea how to cope with this problem, it is faced with the need for another reallocation at the moment[39] and reallocation promises to be one of its most prevalent concerns in the future. According to an FCC report:

> Proceedings involving reallocation of the spectrum can be expected to occur much more frequently in the future. This is particularly true in the most heavily populated portion of the spectrum between roughly 100 Mhz and 10 Ghz, which is the optimum range for most existing applications of radio. Bandwidth requirements preclude the use of lower frequencies for many of these applications, while the propagation characteristics of electromagnetic energy impose economic constraints at frequencies above this range.[40]

Reallocation forces the FCC to make a comparison between the economic value of the different communications services. Prices, as we have seen, provide an objective standard with which to make these comparisons. But prices can arise only if channels can be traded among private owners.

The need for precise comparisons of the economic value of alternative uses of the frequency spectrum is heightened by recent technological developments. Since the invention of the transistor in 1947 and the integrated circuit in the late 1950s, electronic communications and computer technology have become increasingly integrated. The refinement of electronic communications has made the various technologies— broadcasting, cable, fiber optics, and satellites—close technical substitutes for each other. Where the allocation plan of the 1940s and '50s assumed that over-the-air broadcasts were the only way to get a tele-

[37]See Hatfield, "Remarks Before the SIRSA Annual Membership Meeting, October 10, 1980.

[38]"[T]he demand for channels [in the land mobile services] has been so great that the channels have become overloaded almost as fast as the Commission could make them available." Cornell and Lukasik, p. 21.

[39]Existing microwave services in the 12.2–12.7 Ghz band would be displaced if the FCC authorizes the use of that band by Direct Broadcast Satellites.

[40]Robinson, "An Investigation of Economic Factors in FCC Spectrum Management," p. 8.

vision signal to a home receiver, for example, the new technology provides a broad range of options. In addition to co-axial cable, there are MDS, DBS, and Subscription TV, none of which use the VHF and UHF frequencies allocated to TV by the FCC. And, of course, telephone companies are fully capable of providing cable service as well. Telephone service itself does not have to be confined to the traditional wire connections of the local operating company, but can also be provided by satellite networks, by land-based microwave networks, or by cable "television" franchises.

As a result, *every* communications service that uses radio frequencies is competing with every other for "room" on the spectrum. And every service that uses radio technology may use cable or fiber as a technical substitute. In short, every telecommunications service is competing with every other telecommunications service—or rather, they *could* be, if our regulatory system would allow it. For at the heart of the present system of frequency allocation is the assumption that the different services and technologies of telecommunication are not integrated and do not compete with each other. Indeed, a central feature of any system of administrative planning is that the planners must be able to *classify* and *categorize* the subject of their decision-making.

This prerogative of planning forces the FCC to draw a number of intricate and increasingly tangled lines through telecommunications and computer technology. Private mobile radio services, to use only one of many examples, may not be connected to the wireline telephone system. This restriction, in the words of one authority, "may not serve a purpose, other than to keep the two services legally distinguishable."[41]

It is becoming increasingly obvious that classification of telecommunications technologies in an era of microelectronic integration is pointless. "With the emergence of new multi-functional technologies such as microwave, fiber optics and co-axial cable, classifying a service has become a difficult task," claimed the NTIA in a recent report to the FCC.[42] "The results are not always predictable, and . . . courts have not hesitated to overturn the Commission's service classifications." In some cases, the same service is declared by the courts to be two different things. In 1978 a court ruled that MDS services are "broadcasting" and overruled the Commission's ruling that MDS is a "common carrier."[43]

[41]Webbink, "Frequency Spectrum Deregulation Alternatives," p. 27.

[42]Comments of the NTIA, "Inquiry into the Development of Regulatory Policy," p. 77.

[43]*Ortho-Vision, Inc.* v. *Home Box Office, Inc.*, 474 F. Supp. 672 (S.D.N.Y. 1978).

A year later, another ruling held that MDS was *not* broadcasting and based its decision upon the technology used.[44] The FCC originally classified MDS as a common carrier in 1972, but that was before the FCC itself added enough frequencies to the bandwidth of an MDS channel to make it viable as a carrier of television signals. Since then, MDS has simply joined broadcasting, satellites, and cable as another kind of technology used to distribute a television signal. It is not significantly different from Subscription TV.

Any information—text, TV signals, facsimile, data, radio, or voice—can be carried by cable, satellite, fiber, radio, or by any combination of these elements. Why, then, does the FCC expend time, money, and effort attempting to fit integrated technologies into arbitrary pigeonholes? Because its control of the frequency spectrum, and its general determination to regulate communications, forces it to. There is no way a central planning authority can decide which channels to give to which applicants unless they are broken down by service. These arbitrary service classifications impose a pattern of technological and economic segregation upon an integrated industry.

Although segregation by service makes price comparisons difficult and in some cases impossible, the integration of telecommunications technology means that economic considerations should be paramount in the choice of a communications system. If the different technologies can all do much the same thing, the most important question is which can do it at the least cost to consumers.

The relationship between technical and economic considerations in the choice of a communications system is clarified by the graph in Figure 1, taken from the Mathtech study.[45] The graph shows a curve PQR representing all the technically efficient combinations of radio frequencies and non-radio technical inputs needed to produce a given level of radio service. Technical efficiency refers to producing the maximum output per some unit of input. A television transmitter and receiver with a 1.5 Mhz band, for example, may be more technically efficient than ones with a 6 Mhz band, in that transmission and reception of the same signal requires a smaller bandwidth. But technical efficiency must be differentiated from *economic* efficiency, which means that the desired output is produced at the *lowest cost* in terms of all the

---

[44]*Home Box Office, Inc.* v. *Pay TV of Greater New York, Inc.*, 467 F. Supp. 525 (S.D.N.Y. 1979).

[45]*Economic Techniques for Spectrum Management: Final Report*, p. II-10. Hereafter this report will be referred to as "Mathtech."

FIGURE 1

TECHNICALLY POSSIBLE COMBINATIONS OF SPECTRUM AND OTHER INPUTS TO PRODUCE A GIVEN LEVEL OF RADIO SERVICE

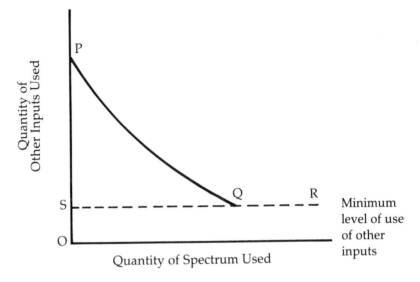

inputs used. The smaller bandwidth may require technical inputs that make the system too expensive and hence economically irrational. Every point on the curve PQR is technically efficient, in that it would be impossible to reproduce the same level of radio service without using more of at least one input, thereby moving either northwest or southeast along the curve. The point is that technology provides us with a range of options, but does not tell us which one is socially efficient.

The vertical axis of the chart represents all those technical inputs that do *not* use radio channels, such as telephone wire, cable, or special attachments designed to reduce bandwidth. The horizontal axis represents the "quantity of spectrum"[46] used. The Mathtech authors point out that under the present system, all the technical inputs on the vertical

[46]Given the concept of the frequency spectrum outlined in section 2, the "quantity" of spectrum is rather a meaningless concept, and the Mathtech study was forced to admit as much (see pp. II-8-9). However, if the horizontal axis is reinterpreted as the extent to which one transmitter is incompatible with another, the meaning of the diagram is unchanged.

86

axis command a price, but the occupation of channels does not.[47] As long as there is no market price for channels, they assert, the designer or user of a communications system has an incentive to locate his system nearer to point Q on the curve—that is, to use radio channels as a cheaper substitute for those telecommunication techniques that are priced. This occupation of scarce channels does, however, displace other users. Those who are displaced are forced to use more expensive inputs or to forgo their communications system altogether.

This chart is valuable for the way it demonstrates that radio communication exists on an economic and technological continuum with other telecommunication techniques. The horizontal axis represents the range of channel dimensions. Under the present system, the FCC's allocation policy determines where on the spectrum a channel will be located and how many there will be, while assignment criteria define the dimensions of each channel. These decisions are of necessity, in the absence of prices, rather arbitrary. In some cases the difficulty of obtaining an assignment drives users to cable or fiber alternatives, while in other cases outmoded allocation and assignment criteria give certain users free access to scarce and valuable channels. Rather than suggesting that the FCC create incentives to use more or less of the spectrum than an efficient market system would, we can conclude with William Meckling that "[FCC] spectrum management is so bizarre that none of us can even imagine what efficient utilization of the frequency spectrum would look like."[48]

A price system would make each user bid against all the alternative uses for a channel. He could not displace other users without paying the price necessary to do so. If his occupation of a channel of specific dimensions is more valuable to him than to the other potential users, then and only then will he retain it. If he can fulfill his communications needs by using alternatives to radio, then the direct price comparison between radio and non-radio alternatives will make it clear when it makes sense to do so. Because he must pay a price to displace other users, a price that increases as the demand for radio channels increases, he is automatically given an incentive to find that point on the curve PQR that minimizes the opportunity cost of using the frequencies. Thus, a price system ensures that an optimal mix of inputs will be selected, a mix that will accommodate as many users as possible at the

[47]Mathtech, pp. II-11-14.

[48]William Meckling, Foreword to *A Property System Approach to the Electromagnetic Spectrum* (Washington, D.C.: Cato Institute, 1980), p. xii.

lowest cost. It follows logically that the most efficient use of the spectrum can be attained only if every communications service and technology is bidding against every other for access. A user in one service not only displaces potential users in the same service, but also limits the number of frequencies available to other services. In other words, the present practice of allocating discrete blocks of frequencies to distinct services is inherently inefficient.[49] Allocation seals off each service into a distinct economic fiefdom, preventing any direct comparison of economic value. This is especially harmful now that many services in different allocated bands, such as DBS and VHF and UHF television, and services which use radio channels for relay purposes such as cable, are competing against each other.

Controlled Markets Are Not the Answer

The case for introducing a price system into frequency allocation is so overwhelming that one is led to wonder why it hasn't happened. The answer is that prices can only emerge from actual market exchanges, and therefore the introduction of a price system requires the definition and free exchange of property rights. Definition of such rights is resisted by Congress and the spectrum-management establishment for a number of reasons.

One of them is simply the government's traditional reluctance to relinquish control of things. But inertia is no justification for the present system. Another reason is the tradition of "public" ownership as it is written into the Communications Act. But "public" ownership, as we have seen, has always been something of a legal fiction, and the present deregulatory trends make it even more of one. A market property structure would allow open competition and create incentives to use radio channels more intensively. This would lead to a greater variety of telecommunication services at a lower price. By removing restrictions from broadcasters and other services without creating such a property structure, Congress is giving us every kind of deregulation except the kind that would benefit the public most.

By far the most important obstacle, however, is the lingering belief that we can have our cake and eat it too. The spectrum-management establishment is more favorably disposed toward administered auctions, lotteries, user fees, and other "controlled market" techniques than freely transferable rights, for the obvious reason that these milder

---

[49]Block allocation was also criticized as technically inefficient by the Joint Technical Advisory Committee report, *Spectrum Engineering, the Key to Progress* (see pp. 76–77).

measures would preserve their control of the spectrum while making the existing system more rational. Most proposals for the reform of frequency allocation are confined to these halfway measures.

Once again, the modern controversy closely follows the debate over economic planning held decades ago. After Mises had called attention to the impossibility of economic calculation under central planning, socialist intellectuals attempted to counter his argument by showing how administrative price-setting and trial-and-error techniques could simulate market prices. Contemporary spectrum managers and economists, however, show no evidence that they are aware of this debate, one of the most important in the history of economic thought.[50]

Interpretations of the debate vary, but to this writer the exchange clearly established that meaningful prices can only emerge from actual exchanges of property in the market. A central planner's disposal of a resource will cause him no direct gain or loss, nor does he directly use the resource or the money exchanged for it. Thus, a shadow price invented by a central planner tells us nothing about the value of the resource in alternative uses. As Hayek has stressed, market competition is a "discovery procedure."[51] Alternative uses and combinations of goods are tested by rival entrepreneurs. These alternatives are "tested" in the strict, empirical sense in which scientific hypotheses are tested in experiments. Just as we cannot say that we *know* the outcome of an experiment until it has been actually conducted, so we cannot know what is the most efficient arrangement of resources unless private owners are free to exchange them in whatever way they see fit in an effort to find out what works best. Prices are the exchange-ratios that emerge from this discovery process. Unless a scarce good is controlled by owners who are free to buy, sell, subdivide, or reconstitute portions of it, its value in alternative uses can never be expressed by prices.

The difference between true prices and administratively set prices can be clarified by analysis of Figure 2, which symbolizes a market

---

[50]An account of this debate, and criticism of the inaccuracy of most standard accounts, is provided by Don Lavoie, "A Critique of the Standard Account of the Socialist Calculation Debate," Ph.D. dissertation, New York University, to be published in 1984 by Cambridge University Press. See also Trygve G. B. Hoff, *Economic Calculation in the Socialist Society* (London: Wm. Hodge & Co., 1949; reissued in 1981 by Liberty Press, Indianapolis).

[51]"[W]herever the use of competition can be rationally justified, it is because we do *not* know in advance the facts that determine the actions of the competitors. In sports or in examinations, no less than in the awards of government contracts or of prizes for poetry, it would clearly be pointless to arrange for competition if we were certain beforehand who would do best." "Competition as a Discovery Procedure," from *New Studies in Politics, Economics and the History of Ideas* (Chicago: University of Chicago Press, 1978), p. 179.

FIGURE 2

HYPOTHETICAL COST-SPECTRUM TRADE-OFF CURVES

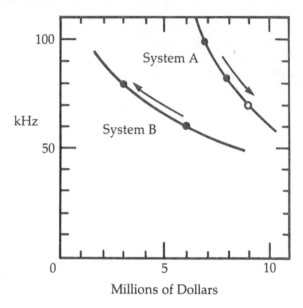

Millions of Dollars

exchange between two owners of radio communication systems.[52] The chart shows two hypothetical cost-spectrum trade-off curves for communication systems A and B. System A operates with a 100 khz bandwidth at a cost of $7 million while System B operates with a 60 khz bandwidth at a cost of $6 million. Assuming that neither system's capacity or quality would be jeopardized, Figure 2 reveals that an exchange of frequencies would be in the interest of both system owners. The owner of System B could pay the owner of System A $2 million for the right to expand his bandwidth by 20 khz. This exchange would move B's system northwest along the curve, reducing the cost of his system by $3 million. After the trade, each system realizes savings of $1 million, while the total number of frequencies occupied by both systems together does not change. As Ewing points out, similar examples could be constructed to show how market exchanges could create greater communications capacity or use less bandwidth while remaining at a fixed total cost.

This chart demonstrates several important features of prices that

[52]Ewing, "Controlled Markets for Spectrum Management," p. 14.

emerge from actual market exchanges. To begin with, the transaction establishes a precise correspondence between a specific act of technical adjustment in the two systems and a specific quantity of money. Economic and engineering considerations are automatically integrated; indeed, there is no way to separate the two, for the price of the extra 20 khz actually *determines* the bandwidth of the two channels. An administratively determined price, in contrast, is established independently of any specific communications system and independently of any exchange of frequencies. Setting a price and adjusting the engineering specifications (e.g., channel or allocation size) are inherently separate and distinct acts. The administrator *hopes* that the price he sets will result in the kind of efficient redistribution of frequencies symbolized by Figure 2, but he cannot guarantee that it will, and in many cases it is likely that the market will be steered in a direction entirely different from what he intended. Had an administrator decreed that 20 khz was "worth" $4 million, for example, the exchange between A and B never could have taken place.

The exchange between A and B was predicated on a purely subjective factor: the judgment that no degradation of service quality would result. But frequencies cannot be plugged into or wrenched out of a communications system without affecting its performance under some conditions. Clearly, only the *users* of the systems can decide whether its integrity would remain intact after the exchange, and the judgment they made would depend entirely on their unique desires and purposes. A mobile radio service for delivery trucks needn't be as worried about signal quality as a High Definition Television broadcaster. Also, in a free market the owners of both systems may be confronted with other opportunities for exchange that would make the symbolized exchange, though attractive in the abstract, undesirable in fact. A, for example, may be capable of splitting his channel into two separate systems of 50 khz each; selling one of them might bring in more revenue than B's offer.

In conclusion, 20 khz is worth $2 million *only* given the specific cost-spectrum trade-off curves of Systems A and B, *only* after a subjective determination that the integrity of the two systems would be harmed by the exchange, and *only* assuming there are no better alternatives facing the two owners. This underscores the sense in which true prices can emerge only from the economic choices of individual owners. Prices that emerge from actual exchanges are sensitive to unique conditions and spur systemic or general adaptation to them. Administrative prices do not emerge from concrete conditions; they are abstractions imposed from above. Thus, they are inherently incapable of duplicating the adaptive function of true prices. In effect, they are simply the same old guessing game in disguise.

Like administered prices, government-administered auctions and lotteries are often put forth as techniques that will introduce "economic" factors into spectrum management short of a true, private property-based market system. Channels with multiple applicants could be auctioned off to the highest bidder, and the auction "price" used to adjust frequency allocations. Or the selection could be made in a frankly arbitrary manner, by lottery. Undoubtedly, both techniques would be less costly and time-consuming than comparative hearings. But while auctions and lotteries may make the distribution of channels more efficient from the perspective of the *government,* they do little to make frequency allocation more adaptable to the demands of the *public.*

By holding auctions for channels, the government is not introducing "economic techniques" into spectrum management; it is simply selling its monopoly privilege to the highest bidder. Auctions do not automatically adjust the dimensions of channels or the size of allocations to supply and demand as a true price system would. They merely give the administrator more information with which to change his allocation and assignment criteria in the traditional way. Auctions can take place only within a service allocation already designated by the FCC, and the bidding must be confined to channels whose dimensions are already defined by the FCC. A high auction price in one allocation may indicate the need for more frequencies in that band—but it does not tell the FCC where to get more frequencies, nor does it provide any information about how many new channels should be created. These adjustments must rely on the guesses of administrators just as the present system does. Moreover, once the adjustments are made the administrator doesn't know the value of scarce frequencies under this new arrangement until and unless he holds another auction. In a true market price system, in contrast, exchanges—and hence adjustments—can be made at any time.

By limiting the scope of market forces to a narrow band of frequencies and a fixed point in time, the FCC would favor large commercial bidders. As one authority noted, "competitive bidding for scarce spectrum under rigid constraints on short-run supply cannot but generate very high market clearing prices."[53] For all their problems, auctions are not that much more flexible than the present hearings system. They still

---

[53]Levin, *The Invisible Resource,* p. 152.

require time-consuming bureaucratic procedures to decide when and under what conditions to hold them, and they can take as long as six months to administer.

Lotteries are not an "economic technique" but merely an administrative expedient. They, too, fail to introduce price signals and incentives into the actual process of frequency allocation. Even the advocates of lotteries admit that unless the lucky person to win one is free to exchange his license with users who may value it more than he does, lotteries fail to guarantee optimal spectrum use.[54] If the value of lotteries is predicated on the possibility of the ensuing market exchanges, it is hard to see why we should bother with them at all. Why not just allow free transferability to begin with?

It is predictable that the spectrum-management establishment would be predisposed to find a middle ground between a pure market and the present system. But there is no middle ground. Either scarcity in radio communication is rationed by a price system—that is, by individual owners of radio transmitters and receivers making exchanges—or it is rationed by the government. The government may streamline its planning process in an effort to make its allocation more responsive to supply and demand, but these efforts should not be misrepresented as the introduction of prices and markets. There is no market in radio communication unless there are freely transferable rights.

---

[54]Webbink, "Frequency Spectrum Deregulation Alternatives," p. 33.

IV. A System of Freely Transferable Rights

With the desirability of private property rights in radio firmly established, the question that remains is how such rights can be defined and put into practical use. The problem is not so much the *definition* of rights—the present system already does that. The problem is to come up with a definition that will hold up throughout the process of market *exchange*, rights that do not rely upon the existence of a central authority for their distribution.

In June of 1969 a team of economists, engineers, and attorneys published a detailed description of an alternative property system in the *Stanford Law Review*.[55] Most discussions and criticisms of freely transferable rights in radio use this study (hereafter referred to as the "DeVany system") as a bench mark. In the DeVany system, rights consist of a geographic area outside which the field strength of a radio signal cannot exceed a specified limit and within which no other transmitter's emissions can exceed the same limit. Rights would also include a specified frequency band. Outside that band the right-holder could not exceed a certain field strength (expressed in volts/meter/hz). Within that band, no other transmitter could exceed the same limit. The DeVany system is the same in all essential respects as the property proposal of Jora Minasian.[56] Instead of controlling the dimensions of the property right by centrally specifying the inputs of the transmitter like the present system, the DeVany/Minasian proposals set limits on out-of-band and out-of-area emissions.[57]

[55]Arthur S. DeVany, Ross D. Eckert, Charles J. Meyers, Donald J. O'Hara, Richard C. Scott, "A Property System Approach to the Electromagnetic Spectrum," *Stanford Law Review* 21 (1969): 1499–1561. Reissued in 1980 by the Cato Institute.

[56]Minasian, "Property Rights in Radiation: An Alternative Approach to Radio Frequency Allocation," *Journal of Law and Economics* 18 (April 1975): 221–272. Note that Minasian considers the rights involved to be rights of "radiation" rather than rights to communication.

[57]Because the present system bases rights on input specifications and the DeVany system uses output specifications, the debate that followed centered on the relative merits of "inputs" vs. "outputs" as a means of specifying rights. It is generally assumed that there

In section 2, scarcity in radio was analyzed as a product of the technology of radio communication. No ethereal natural resource is needed to account for it. The significance of what may appear to be a purely semantic problem will now become evident. The assumption that the spectrum is a "thing" or resource independent of the trans-

---

is a significant difference between the two methods. But there may be less difference than we think.

Describing the definition of rights under the present system, Cornell and Lukasik note that "rights that are defined at this time are input rights granted as privileges by the radio license." But they go on to add that "the feature of a license that is valued . . . is the quality of the output at the receiver location." In fact, input rights make little sense without reference to the signal receiver. Input specifications must be based on interference standards: "The Commission has, in fact, recognized that the ultimate test of a communication system's performance is the quality of the output. In the broadcast and common carrier services, the Commission has established standards for spectrum output in terms of the signal to interference ratio, or noise, to be expected at any receiving location" (p. 13).

If input specifications rely on assumptions about output, the reverse is also true: Any attempt to define rights in terms of output quickly devolves into a set of assumptions concerning input. Under the DeVany and Minasian systems, output limits expressed in volts/meter determine the geographic area owned by a transmitter. Minasian describes this as a set of symmetrical "emission" and "admission" rights (p. 232). But when the actual workings of this system are explored, the need for knowledge of the transmitter's actual inputs becomes evident. In particular, the process of *exchanging* and *enforcing* rights would necessarily involve input specifications as a point of reference. Minasian notes that in purchasing all or part of the rights of a neighboring transmitter, the right-holder "must, to consummate the sale, negotiate with other right-holders to obtain their permission for his increased radiation or nullify the effect by reducing his radiation into the other areas" (p. 238). In practical terms, this means that the right-holder would go to his neighbors with a specific input adjustment to propose to them: I will reduce my power by this much, reduce antenna height by thus much, and so on. Transactions, it seems clear, would have to be defined in terms of input adjustments so that the parties to the transaction and their neighbors could calculate the propagation pattern and field strengths resulting from the exchange. While these exchanges would be *constrained* by the output limits, the actual point of negotiation would have to be inputs. Output limits cannot be exchanged.

The DeVany/Minasian system's reliance on input specifications emerges even more clearly when enforcement of rights is considered. Clearly, it would be too expensive to continuously monitor the entire output pattern of every radio transmitter. Thus, Minasian says that an output-based system would have to rely on occasional monitoring. If measurements led to suspicion that a transmitter was exceeding its boundaries, the *inputs* of the transmitter could be checked to confirm whether a violation was taking place (p. 255). Because of the variability of signals, a measurement of excessive field strength is not proof of wrongdoing; likewise, determination of inputs is not necessarily a rights violation unless it can be shown that the inputs selected will result in an excessive field strength outside the area rights of the transmitter in a significant number of cases.

In conclusion, input specifications have little meaning without reference to a desired signal/interference ratio in a specific receiver, and output specifications have little meaning without reference to the inputs of the transmitter. It does seem clear, however, that knowledge of the actual inputs used by a transmitter is an indispensable part of defining, exchanging, and enforcing rights.

mitter and receiver naturally leads to a search for ways to divide that "thing" into parcels that can be bought and sold by private owners. If the transmitter and receiver hardware and inputs themselves are not the "property" that is to be bought and sold, then all we are left with is the *propagation pattern of a radio signal.* Thus, the attempt to define freely transferable rights becomes a problem of defining and enforcing *boundaries* between these propagation patterns. In theory, the DeVany/ Minasian proposals solve this problem by specifying an absolute limit on field strength in the frequency and geographic dimensions of the propagation pattern. Critics of the DeVany system, however, have raised some serious questions about the practicality of enforcing and exchanging rights without knowledge of the *inputs* used by each transmitter.[58] It is often difficult to monitor the actual output pattern of a transmitter without knowing the antenna height and location, power input, and transmission method.

The main problem with the DeVany/Minasian proposals, however, is that their attempt to base rights on propagation output patterns, while theoretically workable, is unnecessarily complicated. A simpler way to approach the problem is to identify the transmitter and receiver hardware and inputs as the "property" that is owned and traded, while treating interference as a negative *externality* that arises from the use of that property.

"Externalities" is a familiar economic concept. They arise when the consequences of using property in a certain way are not fully visited upon the property owner. It may be that the benefits caused by an owner's use are spread to non-owners (positive externalities) or that the harm caused by the owner's action falls outside the sphere of his legally defined liability (negative externalities). A common example of negative externality is air pollution. A factory pours soot or chemicals into the atmosphere and everyone in the area—not just the factory owner—must breathe the polluted air. (This example makes it clear that the term "negative externality" is often just an antiseptic synonym for invasion or aggression.) If the factory is taxed for its pollution or held liable for the damage it causes, then the externalities are internalized to some degree. A positive externality occurs, for example, when someone lavishly landscapes his property. All of his neighbors benefit from the improved view, and their property values may even rise, but the owner reaps no economic benefit from them.

A statement in Ronald Coase's seminal article "The Federal Com-

---

[58]Levin, *The Invisible Resource,* pp. 94–95.

munications Commission" strongly suggests the connection between the economic concept of externalities and the definition of property rights in radio. "Every regular wave motion," Coase noted, "may be described as a frequency. The various musical notes correspond to frequencies in sound waves. The various colors correspond to frequencies in light waves. But it has not been thought necessary to allocate to different persons or to create property rights in the notes of the musical scale or the colors of the rainbow. To handle the problem arising because one person's use of a sound or light wave may have effects upon others, we establish the right which people have to make sounds which others may hear or to do things which others may see." Coase concluded that "what is being allocated by the FCC or, if there were a market, what would be sold, is the right to use a piece of equipment to transmit signals in a certain way."[59]

In effect, Coase was suggesting that interference be considered a negative externality. The definition of property rights in "the spectrum" *per se* struck him to be as outlandish as creating rights in the colors of the rainbow. The property—the object that is owned—is the transmitter itself, while the owner's property *rights* are limited in accordance with the effects its use will have upon other property owners. Those who followed up on Coase's pioneering call for the definition of property rights in radio, however, abandoned this view. The detailed expositions that followed all attempted to base rights on the signal's propagation pattern. The problems inherent in monitoring something as variable as a propagation pattern have in turn led to a widespread belief that definition of private, freely transferable rights is either impossible or so complicated and expensive to enforce as to be impractical.

The problem is not that property rights cannot be defined, but that we are looking for them in the wrong place. Coase was on the right track to begin with. Properly understood, property in radio is not much different from property in any other material good. Rights in radio, as Coase suggested, consist of "the right to use a piece of equipment to transmit signals in a certain way." More specifically, it would involve the right to place a radio transmitter in a certain location and use a specified bandwidth, power, antenna height, and transmission method, just as the FCC license stipulates now. Under a system of freely transferable rights, however, transmitter (and receiver) owners would be able to subdivide, alter, or reconstitute these rights in any way that did not interfere with the established receivers of those not a party to the

---

[59]Coase, "The Federal Communications Commission," pp. 32–33.

exchange. New entrants could select and use any location and inputs that did not cause harmful interference to established receivers. Once a transmitter's inputs were registered with a central clearinghouse, the channel to specific receivers established by those inputs would be protected from interference by law.

It is a common assumption that the definition of rights in radio is a much more complicated affair than the definition of rights in land or other material goods. But once the *externalities* of these different kinds of property are taken into account, rights in radio seem much simpler. Property in land is plagued by a host of complex externalities, from noise pollution to public health problems. In contrast, the mathematical nature of electromagnetic wave propagation gives us the ability to calculate and predict the externalities caused by the use of radio transmitters. Our knowledge of radio propagation is far from perfect, but it certainly exceeds our knowledge of other kinds of externalities. Who is liable when a noisy bar disturbs the apartment dwellers across the street; when the personal hygiene habits of one household lead to disease or discomfort in another; when a chemical company stores its waste on property that is later sold, and after 10 years the chemicals begin to enter the water table? Where does one set of property rights begin and the other end? In an attempt to resolve such questions, Western societies have evolved complicated institutional arrangements for making judgments about permissible use of property. This machinery encompasses federal, state, and local court systems and regulatory bureaucracies, and other forms of government.

Externalities in radio communication are absurdly simple by comparison. Given the inputs of a radio transmitter, its propagation pattern can be modelled using topographical data and propagation theory, and the model tested by measurements in the field. Variability due to environmental changes can be accounted for by using estimates of probability: For example, at location L the field strength of transmitter T will exceed x microvolts/meter 50 percent of the time. Given this information, a radio transmitter owner knows more about how to avoid negative externalities to other transmitters' receivers than city planners, apartment building owners, developers, and tenants know about how to avoid negative externalities among themselves. The problem of defining property boundaries, then, becomes a problem of coordination or of avoiding harmful externalities to the receivers of other transmitters.

The best thing that can be said for a system of property rights based on this principle is that it is not a purely theoretical construct. A modest version of it called *frequency coordination* is already in use in the 4–6 Ghz band. There is also an historical precedent. In October of 1926 the Tribune Company, publisher of the *Chicago Tribune* and owner of Station WGN, brought a complaint against the Oak Leaves Broadcasting Company in the circuit court of Cook County, Illinois.[60] WGN had been broadcasting at 990 khz since 1923. In September of 1926 the Oak Leaves Company, which had previously been broadcasting at 1200 khz, shifted its frequency to somewhere between 990 and 950 khz, interfering with reception of WGN. As both stations were licensed by the Department of Commerce, as required by the Radio Act of 1912, the conflict had to be settled by the state court. In effect, the Tribune Company argued that it had a property right to its channel, while the defendants insisted that a wave length could not be made the subject of private control. The judge resolved the case in favor of the Tribune Company. One of the major issues was how far the two transmitters should be separated by frequency. The judge upheld the Tribune Company's contention that separation by 40 khz or less would cause harmful interference to WGN: "The court feels that a distance removed 50 kilocycles from the wave-length of the complainant would be a safe distance. . . ." He also enjoined the Oak Leaves Company from causing any material interference to receivers within a 100-mile radius of the Tribune station.

The Oak Leaves case provides us with one of the earliest examples of the establishment of property rights through frequency coordination. The judge's establishment of the proper separation among transmitters in space and frequency, although accomplished without the aid of the sophisticated monitoring devices available today, was similar to the process of frequency coordination as it is practiced in the microwave band today.

The Mathtech study provides a detailed explanation of how frequency coordination works in the 4–6 Ghz band and how its application can be extended to other services.[61] The following is only a summary of this work.

Whenever a new communication system is contemplated, say a satellite link-up, worst-case technical calculations are made to locate the "coordination area," i.e., the area in which the *potential* for harmful interference exists. These worst-case calculations rely on standards set by the International Radio Regulations. Any proposed new stations

---

[60]*Tribune Company* v. *Oak Leaves Broadcasting Station,* Circuit Court of Cook County, Ill. Reprinted in *Congressional Record,* Senate, December 1926, pp. 216–219.

[61]Mathtech, chaps. V and VI, pp. I-6 through I-9.

must communicate the technical details of the proposed station to all existing users within the coordination area, determine whether interference will be caused, and obtain the agreement of other stations.

If interference is estimated to be a problem, bargaining over possible adjustments to the inputs of the existing and proposed stations ensues. The proposed station can abandon its proposed site and move to another; restrict the direction in which its earth station points; construct physical barriers to interference such as pits, embankments, and metallic shielding; or use electronic interference cancellers. Alternatively, the station can pay existing users to install a new, more heavily shielded antenna, a more directional antenna, or pay for the purchase, installation, and maintenance of an interference canceller at an existing station. It can also pay for a change in their frequency.

As the Mathtech authors make clear,[62] frequency coordination is used to define and transfer property rights. The FCC does not establish a rigid assignment table, but allows the users themselves to define the allowable limits of interference and the adjustments that must be made by newcomers. Moreover, the system works. Interference is virtually nonexistent—if anything, the system is too conservative—and the process costs the taxpayers nothing. Private frequency coordination firms such as Compucon and Spectrum Planning, Inc. make money by doing the coordination for smaller firms; larger corporations like AT&T, Comsat, and Western Union do their own frequency coordination. One of the common speculative criticisms of a system of freely transferable rights in radio is that transaction costs, the costs involved in negotiating exchanges of property rights, would be too high.[63] But our actual experience with free transferability refutes this contention. The fact is that businesses engaged in radio communication have a strong incentive to find ways to minimize transaction costs.

The small band of frequencies in which frequency coordination prevails is subject to rapid growth. In 1979 there were more than 16,000 radio relay stations and more than 1,800 earth stations using 4–6 Ghz. Thousands of new receive-only earth stations have been installed recently to serve cable television systems. As an entry procedure, frequency

---

[62]Mathtech, p. V-21.

[63]"No means have been determined by which to define property rights in the spectrum that are consistent with free market operation yet enforceable at reasonable cost." Robinson is referring to property rights based on output, however. If property based on inputs, i.e., frequency coordination, is considered, his statement is demonstrably wrong. Robinson, "An Investigation of Economic Factors," p. 14.

coordination exhibits the flexibility that is so sorely needed in this era of rapid growth in telecommunications.

It is clear that transmitter inputs provide a firm basis for market exchanges. While the area "covered" by a radio signal varies, inputs remain constant. They are also fungible; that is, they can be divided into homogeneous units suitable for exchange. In return for compensation, a transmitter can agree to raise or lower his power level, to increase or decrease his antenna height, to change antenna location or polarization, and reduce or enlarge bandwidth by a specified number of units. Such a system of freely transferable rights need not engage in a futile search for boundaries in the atmosphere; it need only ask whether a given set of inputs establishes a channel between the transmitter and the desired receiver(s) of sufficient quality for the purpose at hand. With respect to other transmitters, we need only ask whether the inputs selected interfere with their established connections or not. If they do, the inputs of the interfering party must be altered until the interference is eliminated, or the interfering party must offer the other party enough compensation to make the interference acceptable.

The problem of extensive and unpredictable patterns of interference caused by natural phenomena beyond the user's control is inevitably raised as an argument against a system of freely transferable rights. But if rights are based on inputs rather than propagation patterns, natural phenomena pose no problem. If T_1's duly owned inputs suddenly create interference with T_2's channel due to some electromagnetic fluke, then the market, knowing that T_1's owner has established a prior *right* to those inputs regardless of the unpredicted natural phenomena, can adjust in any of the following ways:

T_2's owner can bargain with T_1's owner to induce him to adjust his inputs; transmitters can purchase insurance against such events, either in the form of damage claims or in the form of emergency back-up channels; the channel interfered with, if the problem becomes recurrent, will become devalued just as land subject to flooding becomes devalued, and its price will fall. This creates an incentive to either find a technological solution to the problem or, if none can be found, to transfer ownership of T_2's channel to a use that would not be significantly affected by occasional interference. In either case the market would reapportion the use of frequencies in a rational way.

It must also be stressed that *no* system of property, public or private, can protect people from interference that is truly *unpredictable* in nature. The proper standard of judgment here is not which system can avoid all unpredicted problems, but which can best adjust to them after they happen. A private property system would allow the individuals directly affected by the problem to work out a solution. The introduction of a price system would yield knowledge of the efficiency of various solutions to the problem. The knowledge generated by market transactions would, over time, lead to adjustments in radio usage that would minimize the risk of interference problems. A centralized system lacks this flexibility. When problems arise, the ownership pattern cannot change readily. The FCC can order very conservative input specifications in an attempt to minimize the risk in advance, but this conservatism may needlessly deny hundreds of users the chance to engage in radio communication.

Under a system of frequency coordination, the geographic extension of the property right is determined by the *receiver*. That is, the property right only protects the channels established to specific receivers. It does not give the transmitter a right to exclude other transmitters from the entire geographic area in which his signal exceeds a certain field strength. This is an important characteristic of the system and is explored in more detail in the Appendix. Involving the receiver in this way, however, would work only for "point-to-point" radio communication services. The distinguishing feature of the point-to-point services is that every transmission is *intended* for a specific receiver or group of receivers. There is little difference between the people who transmit and those who receive in these services. Both have a commercial stake in the process. Many times the equipment used is both a transmitter and a receiver, as in microwave relays, satellites, and the mobile radio "community repeater," the antenna that relays the signal from one mobile unit to another.

This is a far cry from broadcasting. The broadcaster is not interested in any specific receiver; he merely wants to cover a geographic region containing a large population with a signal strong enough to allow that population to tune into his programming at will. Likewise, the broadcast receiver is not interested in any specific transmitter *per se;* he merely wants a range of program choices that suits his fancy. Thus, in addition to input specifications, property rights in broadcasting must specify the *service area* of the station; that is, the geographic region within which the transmitter is protected from interference. The DeVany/Minasian proposals could serve as a model here. Once input specifications are included in the definition of rights all of the objections to the feasibility of their property system become invalid. In addition, the Mathtech study provides a detailed explanation of how to apply frequency coordination to FM radio broadcasting. Mathtech would define the service area of the FM station according, to the FCC's "50/50 rule." This is simply the area in which a field strength of 1 microvolt/meter is exceeded 50 percent of the time at 50 percent of the locations measured. With its service area so defined, a proposed FM station would have to determine whether it overlapped the service area of any other station. If so, it would have to obtain (or bargain for) the other's consent before it could begin transmission. If not, it would be free to apply for an FCC license. This is how the Mathtech authors proposed to solve the problem; the important thing is not this particular rule but that some agreed-upon rule would be evolved. As long as both input and geographical rights are freely transferable, the system would be able to adapt and the rule itself could be improved over time.

In sum, frequency coordination is simply a method or procedure for selecting transmitter inputs that do not interfere with established channels. Such a procedure is all that is needed to establish a free market in radio communication. However, the system as it is defined by the Mathtech study is much too conservative. The following modifications are recommended:

● Eliminate all service allocations. The Mathtech recommendations still confine the use of frequency coordination to services already designated by the FCC. As we have seen, however, one of the biggest problems now facing the FCC is the necessity for reallocation. Indeed, the most important function of a freely transferable rights system is its ability to allow new entrants into services where demand is rising, and to allow channel users in services where demand is decreasing to shift their channels to other uses. As long as service allocations are made by a central planning authority, the most beneficial effect of market forces will never be realized.

● Allow open entry. There is no reason to give the potential competitors of a new service—that is, existing transmitters—unilateral power to determine whether a new entrant is acceptable or not. Thus the present practice of frequency coordination, in which new entrants must obtain the prior agreement of all existing users in an area, should be modified. New stations should be allowed to enter at will, the only requirement being that they register their inputs beforehand. Any interference they caused would become actionable only after some demonstrable problem occurred. Existing transmitters should be able to obtain an injunction against a new entrant only in extreme cases; i.e., when there is strong evidence that the new entrant would seriously disrupt their service. Interference problems that emerged after a period of time could be settled on the basis of temporal priority; if transmitter

B started regularly interfering with transmitter A's receivers after a year, B's owner would be liable for the interference if his inputs were registered after A's. The same principle would hold in disputes over intermodulation interference.[64] If A and B together cause interference with C's receivers, and the rights of the owners of both A and C are prior to B's, then B's owner would be liable for the intermodulation problem.

- Register input. All radio transmitters in the U.S. would record their inputs in a central registry, much as county courthouses serve as registries for property deeds. This registry could sustain itself by charging enough of a fee for its registration service to cover its expenses and enforcement costs. The task of monitoring interference and of identifying its source would become the responsibility of the rights-holder. Transmitter inputs not registered would not be protected as property rights and would be revocable if they began to interfere with registered rights.

- Vest rights in existing users. One advantage of a system of freely transferable rights based on inputs is that there would be few transition problems. We need only vest in all existing licensees the right to use the inputs granted to them by the current FCC license. Since they already possess these rights, no defender of the present system can complain that vesting them would be unfair or chaotic. However, some exceptions to vestment may be desirable. The U.S. military, for example, has been granted a huge portion of the spectrum. It is impossible to determine whether it actually needs all of its channels because the cloak of "national security" conceals its requirements from detailed scrutiny. The military must purchase the arms and equipment it needs and must pay salaries to its officers and enlisted personnel. With practically every other scarce good, the military must justify its needs to the Congress. Radio communication rights, in contrast, are granted free. This practice invites waste and prevents the civilian authorities

---

[64]Intermodulation is a specific kind of interference in which the transmissions of two transmitters in close proximity combine to interfere with a third transmitter. Thus, if A is transmitting at 50 Mhz and nearby B is transmitting at 100 Mhz, interference to C, who transmits at 150 Mhz—the *addition* of A and B's frequencies—may result. For some reason, intermodulation is frequently cited as an immense, well-nigh insoluble problem for a system of private property, because neither A nor B alone is responsible for the interference. However, the problem can be handled by a simple rule of priority. If, for example, A and C are transmitting without interference, and it is the *addition* of B's transmissions which causes the problem, then B is liable. If both A and B register their inputs at the same time, then *both* could be enjoined from transmitting until they worked out some kind of coordination among themselves that did not interfere with C.

from understanding the actual value of the portion of the spectrum controlled by the military. For this reason it is advisable to divest the military of its frequencies and force it to repurchase its channels on the market.

Another exception to vestment might be areas, such as television broadcasting, in which the FCC has fostered monopoly or concentration in the past. Thus, while the new low-power television stations may create interference in the outer margins of the signal contour of established stations, the advantages of creating many new channels far outweigh any claim the established broadcasters might have. As long as rights are freely transferable, those LPTV stations that prove not to be viable can transfer their rights to other uses.

- Define a homesteading principle. Legislation defining an orderly process by which individuals could acquire rights to unused portions of the frequency spectrum (e.g., channels above 40 Ghz or channels released by the military) would have to be written.[65]

The creation of property rights in channels naturally implies that there would be no restrictions on the kind of information carried by a channel. This, and the absence of rigid service allocations, would promote competition. If the pay television business carried by most MDS channels slacked off in the future because of DBS or cable competition, the owner could shift his channel to data transmission services or any of an infinite number of other uses. Competition would flourish if channel owners could enter any telecommunications market for which there was adequate demand without restriction or delay. At present, for example, banks that need to exchange data with their branches and with other banks have essentially two alternatives: hook up to the phone lines (and pay enormous monthly bills to AT&T) or build their own microwave relay system. Under a system of freely transferable rights, it may prove more economical to purchase or lease existing channels from other services.

---

[65]Homesteading or initial rights acquisition raises complex issues of *justice* as well as economic efficiency. For that reason this study does not explore the issue or make any proposals. The approach to initial rights definition taken by Coase and Demsetz, however, is clearly inadequate. They contend that the initial distribution or definition of rights is irrelevant; as long as free exchange is possible, rights will move to their optimal arrangement. If a railroad dumps soot onto the crops of a farmer, for example, the Chicago school holds that it makes no difference whether the farmer has to compensate the railroad to stop the pollution, or whether the railroad is taxed or sued to stop the pollution. It certainly makes a difference to the farmer and the railroad.

106

V. The Problem of Regulatory Obsolescence

Telecommunications law is in a state of turmoil. A legal and regulatory system more than 50 years old has had to contend with an explosion of new technologies and new services. The obsolescence of the present regulatory framework is obvious to everyone. But there is no consensus on how to change it.

Attempts to rewrite the Communications Act have not fared much better than the original act itself. The most ambitious of these attempts was initiated in 1977 by former Rep. Lionel van Deerlin. Van Deerlin's systematic overhaul of the act, in its various incarnations as H.R. 13015 and H.R. 3333, was a modest move in the direction of deregulation. For the most part, however, it stayed squarely within the framework established by the original act, and it increased government involvement in some areas. Despite the modesty of its reforms, the bill never made it out of committee. Telecommunications proved to be too complicated and the process of reform beset by too many conflicting special interests. In other words, even though everyone agrees that the Communications Act is in drastic need of systematic revision, it simply cannot be done.

In the 97th Congress, efforts to reform the law were broken down into smaller, more manageable bills. More than seven bills were submitted to the subcommittees covering radio broadcasting, television broadcasting, telecommunications, international telecommunications, and other areas. But a piecemeal approach has its pitfalls, as well. Congress can never be sure just how the pieces fit together. The recent AT&T antitrust settlement, which rendered the meticulously prepared new telecommunications law (S. 898) obsolete virtually overnight, makes it clear that the same fate that befell the original Communications Act and the van Deerlin alternative may yet be in store for these attempts at piecemeal revision.

The specter of obsolescence has dogged telecommunications legislation from the beginning. The Radio Act of 1912, written with ship-shore communications in mind, worked fine as long as the use of telecommunication technology was confined to those purposes. But as soon as a new technology, broadcasting, developed in the early 1920s, the established regulatory system fell apart. With the Radio Act of 1927 and the Communications Act of 1934, the regulators caught up with the industry once again. But the regulatory framework established inevitably reflected the state of technology at that time. The telecommunications industry was segregated by technology and service. Local broadcasters were considered to be something quite different from telephone companies; telephone companies were considered to be something separate and distinct from national broadcasting networks and mobile radio services. Computer companies were, until recently, not directly related to telecommunications companies, and vice-versa. What we have been witnessing in the last 20 years is a breakdown in this regulatory scheme caused by microelectronic integration and the addition of satellites to the field of relay services. Of course, technology continues to evolve, and there is little doubt that a regulatory scheme based on today's conditions will become obsolete in a decade or two— maybe less.

The assumption underlying the original Communications Act, and most contemporary efforts to reform it, is that government should actively shape the development of telecommunications through positive intervention. This assumption is clearly the source of the recurring problem of regulatory obsolescence. Laws are supposed to establish the *general rules* governing human conduct.[66] The process of legislation is slow and inevitably involves a consensus among groups with varied and often conflicting interests. Consequently, legislation cannot provide an adequate basis for detailed control of economic and technological affairs, particularly in an industry as volatile as telecommunications. The proper function of law is to define general and enduring rules of just human interaction, rules that will hold fast through the maelstrom of technological and economic change.

There are two ways, then, in which the revision of the Communications Act can be approached. First, the Congress can stay within the framework of public control and central planning established by the original act. It can modify and meliorate that control, even shed much of it, but as long as that framework remains, Congress will be committed to constant revision and reform as conditions change. It will therefore be tugged back and forth constantly by special interests seeking

[66]Hayek, *Law, Legislation and Liberty*, vol. 1, *Rules and Order* (Chicago: The University of Chicago Press, 1973).

the most favorable terms of revision. The alternative path is to withdraw altogether from the business of shaping the telecommunications industry—just as Western governments in the 18th century withdrew their control over religious beliefs. Instead, Congress should apply First Amendment and free market principles.

The system of freely transferable rights sketched above is an attempt to discover and apply such principles to radio communication. A system of private property rights, in conjunction with the First Amendment, provides a coherent approach to telecommunications reform. A system of property rights or frequency coordination would make the allocation of radio frequencies responsive to changes in supply and demand; the government would not need to intervene in response to changing conditions. It would also allow the owners of radio channels to devote them to whatever information services proved most attractive; there would be no need for the arduous and arbitrary classifications that currently hinder the industry. Owners would be able to exchange, subdivide, or reconstitute channels to adapt to changing technology, again without need of legislative change or regulatory oversight. Because it does not lead to any particular results but lets market forces determine the future of the industry, a system of property rights eliminates one of the major obstacles to communications law reform: the attempt of myriad special interests to get the government to rig the game in their favor. In sum, by defining a fair and orderly procedure for trading and protecting rights in radio communication, Congress can protect the public's interest in justice and efficiency without attempting to exert detailed control over this dynamically changing field.

Appendix: The Role of Receivers

As noted in the text, implicit in the process of frequency coordination is the fact that the *receivers* of a transmission determine the geographic extension of the property right. This is a sharp departure from most approaches to the problem of defining rights in radio, which attempt to base rights on the propagation of the signal in space. To Minasian, DeVany, and the FCC as well, rights in radio are "rights of radiation," i.e., the right to "cover" an "area" with a radio signal. In the system developed here, it is the ability to make connections with receivers that forms the basis of the right.

To illustrate what this means, let us postulate that T_1 and R_1 are two links in a microwave relay 20 miles apart. For the sake of argument, let us also assume that T_1 and R_1 are owned by different firms. By its reception of T_1's transmission, R_1 establishes a channel of a specified bandwidth and geographic length. Together, the owners of T_1 and R_1 have a right to prevent any other transmitter from using a location and/or inputs that will interfere with that channel. But they do not have the right to protect the "area covered" by T_1's signal; indeed, this "right" would be quite meaningless, as we shall see later. If, as is usually the case, the inputs used by T_1 go significantly beyond that needed merely to establish a channel to R_1, the additional inputs are an externality. New entrants can make room for their transmitters by bidding away these externalities, e.g., by offering the owner of T_1 payments to focus his beam, use a more directional antenna, reduce channel size, or ultimately, to connect by wire. In effect, T_1's owner has "homesteaded" the original propagation pattern, in the sense that he has the right to the inputs originally used to establish a channel to R_1. But he does not have the right to prevent the operation of new transmitters that would not affect his channel to R_1.

Furthermore, as long as T_1 and R_1 are owned by different firms, neither of them owns the channel itself. The owners of T_1 and R_1 only own their transmitter hardware and inputs. The channel is an implicit contract or agreement between them, and either of them can revoke it at will. If T_1's owner chooses to cease transmission, then R_1's owner has no right to force him to continue. While this may seem unexceptionable, reversing the relation seems just as logical: A transmitter does not have the right to command a receiver to accept his transmissions. While equally obvious, this conception of the role of the receiver has radical consequences. It means that if R_1 no longer wishes to receive T_1, then T_1—while retaining his right to operate in the same way, using the same inputs—loses any right to protect his channel to R_1 from interference. The ownership rights of the transmitter do not extend to the receiving equipment unless, of course, the same person owns both the transmitter and the receiver. Thus the desirability or undesirability of interference at any given receiver location can be determined only by the receiver. If another transmitter, T_2, comes along and thoroughly drowns out T_1 in R_1, the owner of R_1 has the right to initiate measures to stop this. But he will do so only if he prefers T_1's transmissions to T_2's. If he doesn't object to T_2's interference, then there is nothing illegal about it. More precisely, if the receiver doesn't object to interference, *then it isn't interference.*

"Interference" is often bandied about as if it were a technical term of great precision. It isn't. Interference is nothing more than a ratio between desired and undesired signals in a specific receiver. There is no more objective definition possible. Scientific measurement can determine that signal A is of field strength *a* volts/meter and signal B is of field strength *b* volts/meter at the point of reception at a given time. But science does not and cannot tell us whether the resulting ratio, *a:b*, impairs reception to an unacceptable degree, nor can it tell us which signal is desired and which is undesired. That is inherently a subjective judgment. If the ratio between *a* and *b* is 10:1, receivers in some radio services may be able to function perfectly well. Other services would require something on the order of 100:1 or 1000:1. Moreover, some receivers may object to a slight addition of noise that would pass by others unnoticed.

From the beginning, property theorists in radio have attempted to base property rights on the "area covered" by a radio signal. But this approach stems from a misapprehension of the problem. There is no spatial dimension to electromagnetic emissions. Like visible light, which is merely a small section of the spectrum around 10^{14} hz, radio signals never reach some point and stop; they merely attenuate until they become undetectable. Radio astronomers receive and interpret electromagnetic emissions from galaxies light-years away. Theoretically, observers in those galaxies could receive and decode all the radio transmissions from this planet. Thus the "area covered" by a radio *signal* is, over time, the full volume of space. And any transmitter on earth could be received simultaneously anywhere in the hemisphere in which it is located, provided that the receivers were sensitive enough, their antennas high enough, and no other transmitters overpowered the signal.

While we cannot speak of the "area covered" by a radio signal, we *can* speak of the area in which it can be *received*. And the geographic boundaries of reception are completely dependent upon the unique position and characteristics of the receiver and transmitter in question and the proximity of other transmitters. Once the problem is understood in this way, the crucial role of the receiver in determining the geographic extension of the property right becomes clear.

The question that remains is whether the choices of receiver owners will directly affect the property structure of radio, or whether the transmitter owners or a central authority will make these choices for them. Ultimately, this is a normative question, not a positive one. By definition, judgments about the acceptability of interference are subjective. The regulation of radio interference, then, boils down to a matter of *whose* subjective preferences will prevail. The standards of science or technical and economic efficiency cannot provide us with an answer to this question. We can answer it only by discussing whose preferences *ought* to prevail.

ABOUT THE AUTHORS

Edwin Diamond is Senior Lecturer in Political Science at Massachusetts Institute of Technology. He has worked as an editor for *Newsweek* and *New York Daily News Tonight*, contributing editor of *New York Magazine*, and reporter for newspapers in Chicago and Washington. He also helped to found the *Washington Journalism Review*. He is a frequent television commentator and has won numerous journalism awards. His books include *The Tin Kazoo: Television, Politics, and the News* (1975); *Good News, Bad News* (1978); and *Sign Off: The Last Days of Television* (1982).

Norman Sandler is a White House correspondent for United Press International. He worked as a political reporter for UPI in the Midwest and holds graduate and undergraduate degrees from Massachusetts Institute of Technology. He has written extensively about telecommunications policy.

Milton Mueller is pursuing graduate studies at the Annenberg School of Communications at the University of Pennsylvania. He has written on telecommunications issues for a variety of publications and is currently working on a book on cable communications. He has been a research fellow of the Institute for Humane Studies and is an associate policy analyst of the Cato Institute.

$840336960

The Cato Institute

The Cato Institute is named for the libertarian pamphlets *Cato's Letters*. Written by John Trenchard and Thomas Gordon, *Cato's Letters* were widely read in the American colonies in the eighteenth century and played a major role in laying the philosophical foundation for the revolution that followed.

The erosion of civil and economic liberties in the modern world has occurred in concert with a widening array of social problems. These disturbing developments have resulted from a failure to examine social problems in terms of the fundamental principles of human dignity, economic welfare, freedom, and justice.

The Cato Institute aims to broaden public policy debate by sponsoring programs designed to assist both the scholar and the concerned layperson in analyzing questions of political economy.

The programs of the Cato Institute include the sponsorship and publication of basic research in social philosophy and public policy; publication of the *Cato Journal*, an interdisciplinary journal of public policy analysis, and *Policy Report*, a monthly economic newsletter; "Byline," a daily public affairs radio program; and an extensive program of symposia, seminars, and conferences.

CATO INSTITUTE
224 Second Street, S.E.
Washington, D.C. 20003